AN INTRODUCTION TO

This textbook is a self-contained introduction to tides that will be useful for courses on tides in oceans and coastal seas at an advanced undergraduate and postgraduate level, and will also serve as the go-to book for researchers and coastal engineers needing information about tides. The material covered includes: a derivation of the tide-generating potential; a systematic overview of the main lunar periodicities; an intuitive explanation of the origin of the main tidal constituents; basic wave models for tidal propagation (e.g., Kelvin waves, the Taylor problem); shallow-water constituents; co-oscillation and resonance; frictional and radiation damping; the vertical structure of tidal currents; and a separate chapter on internal tides, which deals with ocean stratification, propagation of internal tides (vertical modes and characteristics) and their generation. Exercises are provided in each chapter.

THEO GERKEMA is a senior researcher in the Department of Estuarine and Delta Systems at the Royal Netherlands Institute for Sea Research (NIOZ), Yerseke. He graduated as a theoretical physicist from the University of Groningen and studied the generation of internal solitons for his PhD at NIOZ. The early part of his career was devoted to studying internal tides and Coriolis effects. In recent years, his work has centered on coastal dynamics, tides, sediment transport and sea-level variability, involving sea-going observational work and modeling.

"Theo Gerkema connects his expertise in internal wave dynamics and shallow-water processes in one consistent and inspiring presentation of tides in the ocean. The outcome is an excellent textbook, which I will certainly use for my own teaching. And I will consult the book for my own research, as a backup for all things about tides which I should know better."

Hans Burchard,
Leibniz Institute for Baltic Sea Research

"In this book, Professor Gerkema presents one of the most comprehensive treatises on tides currently available. Starting with the fundamental concepts of the tide, we get an extensive introduction to its generation and properties, and why it acts the way it does. The book also contains a chapter on the important internal tide, acting as an energy sink in the deep ocean, away from the tidal influence at coastlines. All of this is presented with a minimum of equations, which makes the book all the more accessible. The book draws extensively on the scientific literature, but presents it in an understandable way that should suit readers from diverse backgrounds and levels. The exercises throughout the chapters really challenge the understanding of the reader and make for excellent additions to the book. More specialized concepts are presented in boxes in the text – a nice feature that allows a reader to focus (or not) on what is being presented. The book has the potential to become the go-to reference for students and scientists alike when it comes to tidal dynamics, because it fills a gap by focusing only on tides and covering all aspects of them."

Mattias Green,
Bangor University

"A nicely balanced physical and mathematical account of the tide-generating forces and their multiple periodicities, followed by an insightful presentation of the resulting tidal currents and some of their effects in coastal seas and in the deep ocean, where internal tides are a major cause of climatically important mixing."

Chris Garrett,
University of Victoria

"Professor Gerkema's book offers a comprehensive review of tides, an ancient subject that still inspires a great deal of modern research. Professor Gerkema outlines the basics of astronomical tidal forcing, tidal analysis, and tidal dynamics in the open and coastal oceans. I appreciate his efforts because it is hard to find all of these fundamentals in one place. Gerkema also includes a nice chapter on the subject of internal tides, a topic of much current research now. I will certainly add Professor Gerkema's text to my office bookshelf."

Brian Arbic,
University of Michigan

AN INTRODUCTION TO TIDES

THEO GERKEMA

CAMBRIDGE
UNIVERSITY PRESS

CAMBRIDGE
UNIVERSITY PRESS

University Printing House, Cambridge CB2 8BS, United Kingdom

One Liberty Plaza, 20th Floor, New York, NY 10006, USA

477 Williamstown Road, Port Melbourne, VIC 3207, Australia

314–321, 3rd Floor, Plot 3, Splendor Forum, Jasola District Centre, New Delhi – 110025, India

79 Anson Road, #06–04/06, Singapore 079906

Cambridge University Press is part of the University of Cambridge.

It furthers the University's mission by disseminating knowledge in the pursuit of education, learning, and research at the highest international levels of excellence.

www.cambridge.org
Information on this title: www.cambridge.org/9781108474269
DOI: 10.1017/9781316998793

First published 2019

Printed in the United Kingdom by TJ International Ltd, Padstow Cornwall

A catalogue record for this publication is available from the British Library.

Library of Congress Cataloging-in-Publication Data
Names: Gerkema, Theo, 1965- author.
Title: An introduction to tides / Theo Gerkema.
Description: Cambridge, United Kingdom ; New York, NY : Cambridge University Press, 2019. | Includes bibliographical references and index.
Identifiers: LCCN 2019007597 | ISBN 9781108474269 (hardback) | ISBN 9781108464055 (paperback)
Subjects: LCSH: Tides.
Classification: LCC GC301.2 .G47 2019 | DDC 551.46/4–dc23
LC record available at https://lccn.loc.gov/2019007597

ISBN 978-1-108-47426-9 Hardback
ISBN 978-1-108-46405-5 Paperback

Additional resources for this publication at www.cambridge.org/gerkema.

Contents

Acknowledgments

This book originates from lecture notes that I wrote while giving a course on tides at Utrecht University in 2015. I thank the students for their perceptive questions and helpful feedback.

I am very grateful to a number of colleagues who took the time and effort to read and comment on several chapters of this book: Huib de Swart, Pieter Roos, Kirstin Schulz and Wim Verkley. Henk Schuttelaars and Riccardo Riva provided feedback on portions of the text. All their criticism and suggestions greatly helped me improve the manuscript. It goes without saying that any remaining errors or unclarities are my sole responsibility. I thank Hans van Haren and Louis Gostiaux for kindly providing figures for this book, Matias Duran Matute and Carola van der Hout for providing data for some figures and my nephew Thomas Meutgeert who skillfully created two figures for Chapter 3.

I am pleased to acknowledge that over many years now, my work on tide-related topics has been supported by the Royal Netherlands Institute for Sea Research (NIOZ). I thank Leonne van der Weegen and Marlies Bruining (library of NIOZ) for tracing down some old and hard-to-find papers.

I am grateful to Matt Lloyd, Zoe Pruce and Samuel Fearnley from Cambridge University Press for their guidance during the editorial process.

Working on this book often brought back into memory my former thesis supervisor Sjef Zimmerman, who sadly passed away last year. I am much indebted for everything I learned from him and dedicate this book to his memory.

Finally, I cannot convey in words – but perhaps in flowers? – my gratefulness to Mariëtte for her inspiration, patience, and wholehearted support during the process of writing this book.

1

Introductory Concepts

1.1 The Nature of Tides

The periodic rise and fall of sea level offers the most visible manifestation of the tides (Figure 1.1). This alternation of high and low waters is of great importance for navigation, coastal safety, and coastal ecology. As a rule, a full cycle of high and low waters takes about half a day; these tides are called *semidiurnal*. However, there are also places with daily periods (*diurnal* tides), and still others where an alternation between the two occurs (*mixed* tides). Figure 1.2 maps the distribution of these periodicities worldwide. The tides are predominantly semidiurnal; diurnal tides occur in isolated patches, such as the Gulf of Mexico, the Caribbean, the Indonesian archipelago, the North Pacific, and around the Pacific Antarctic embayment. In Figure 1.2, the mixed tides are subdivided depending on which of the two – diurnal or semidiurnal – is dominant.

Meanwhile, it is important to realize that the classification of Figure 1.2 offers only a rudimentary distinction, which masks the real complexity of the tidal signal as it is found at any given location. In particular, variations on half-monthly, monthly, and longer timescales are generally present as well. As an aside, we note that the terms "daily" and "monthly" are here to be understood in a loose sense, as an indicative length of time.

Figure 1.2 tells us how often per day high and low waters occur. We now consider their heights, expressed as the *tidal range*: the vertical interval between high and low water levels (Figure 1.3). Again making a rudimentary distinction, we classify the tidal range in three categories: macrotides (ranges exceeding 4 m), mesotides (between 2 and 4 m) and microtides (lower than 2 m). Without exception, macrotides occur in patches that are attached to the continents – an indication that the configuration of the continents acts as one of the organizing principles behind the global pattern of tides.

Figure 1.1 View of an embayment in Castro (Chile) during low tides (left) and high tides (right), with the characteristic stilt houses (*palafitos*), designed to be "tideproof." Photographs by the author.

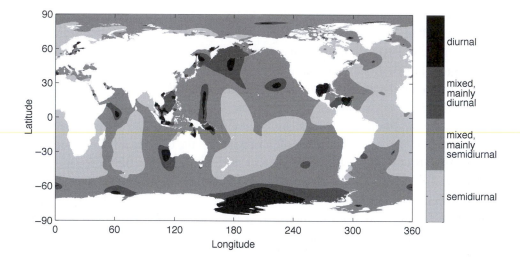

Figure 1.2 Global distribution of semidiurnal, diurnal, and mixed tides. This classification is based on the so-called form factor (to be discussed in Section 4.5), here calculated using data from the global tide model FES2014. This model adopts a data assimilation technique to combine observational data from satellite altimetry and tide gauges with a finite-element numerical tide model. Figure generated using Aviso+ products, courtesy of LEGOS/Noveltis/CNES/CLS.

Figures 1.2 and 1.3 depict the local characteristics of the alternating high and low tides. This way of viewing the tide lies at the basis of tidal predictions, in which a local sea level record is used to predict the moments and heights of future high and low tides at that very location, for example a harbor. The prediction can be made independently of how the tide behaves elsewhere.

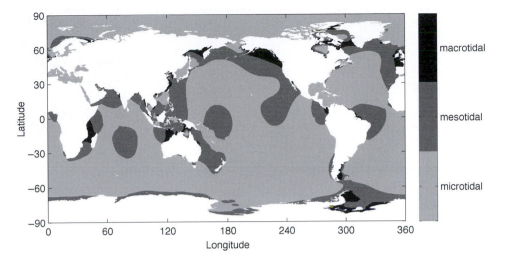

Figure 1.3 Global distribution of the tidal range. Three categories are distinguished: *macrotides:* tidal range > 4 m; *mesotides:* tidal range 2 − 4 m; *microtides:* tidal range < 2 m. The tidal range is here estimated by superposing the (double) amplitudes of the main eight constituents (see Section 4.4 for more details). The amplitudes were obtained from FES2014 model data. Figure generated using Aviso+ products, courtesy of LEGOS/Noveltis/CNES/CLS.

However, from a dynamical perspective, tidal signals at different locations are connected. This can be seen by comparing the moments of high and low tides along a stretch of coastline, as illustrated for the North Sea in Figure 1.4. At a certain point in time, a high tide occurs at the spot marked by a black circle on the east coast of Scotland. Nowhere else along the British North Sea coast do we find a high tide at that same moment. It is only at a distant spot in the southern part of the Netherlands, and then at Denmark, that we find simultaneous high tides. For low tides (gray circles) the spots lie similarly far apart. The distance between locations with simultaneous high (or low) tides typically amounts to several hundred to more than a thousand kilometers. In time, the positions of high and low tides (black and gray circles) propagate along the coast, as loosely indicated by the arrows in Figure 1.4. This presents a glimpse of the tide as a *wave* phenomenon: the high tides being the crests of the wave, the low tides, the troughs. The wavelength, the distance between successive crests, is of the order of hundreds of kilometers or more. This is, anywhere on the globe, much larger than the local water depth; thus, the tide can be classified as a *long wave*. The speed at which high and low tides move along the coast is the phase speed of the tidal wave; it is of the order of 100 kilometers per hour in the example of the North Sea (the phase speed differs per location, though, as it depends on the water depth).

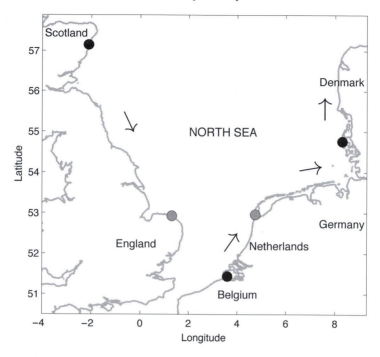

Figure 1.4 A snapshot with locations of high tides (black circles) and low tides (gray circles) along the North Sea coast at a particular instant (based on data from the UK Hydrographic Office). In time, the high and low tides move in a counterclockwise sense along the coast, i.e., southward on the western side and northward on the eastern side of the North Sea, as indicated by the arrows. After one full tidal period, the original situation is replicated.

As with all water waves, a tidal wave as such is an immaterial thing: a signal, an amount of energy that propagates onward. The wave propagation is supported by an oscillatory movement of the water parcels, the *tidal currents*. Tidal currents are of the order of a few centimeters per second in the open ocean, and up to the order of a meter per second in coastal areas. It is important not to confuse the water motion with the wave propagation. In the course of a tidal period, the water particles move for the most part back and forth, while the tidal wave, with its crests and troughs, progresses forward over long stretches. For the water particles, the horizontal distance covered between two successive turnings of the tide is called the *tidal excursion*. It can be expressed as twice the ratio of the tidal current amplitude and the tidal frequency. Its magnitude lies typically in the range of 10–20 km in coastal areas. In the course of a tidal period, particles move back and forth over this distance. In comparison, the vertical movement of the water particles, defined by the tidal range, is much smaller; in other words, the tide plays out primarily in the horizontal. Notice that there is a very clear distinction in scales: the tidal range is much smaller than the tidal excursion, which, in turn, is much smaller than the wavelength of the tide.

The relation between tidal currents and sea level is not a straightforward one. Colloquially, the word "ebb" is sometimes used in reference to the phase of the tidal cycle when the sea level drops, but this confounds the water level with the current. In the strict sense of the word, ebb and flood are about the different phases of the tidal *current*: ebb being the current against the direction of tidal wave propagation or out of a tidal basin or estuary, and conversely for flood. Low and high tides sometimes coincide with maximum ebb and flood, sometimes with moments of slack water (i.e., the turning from ebb to flood or vice versa), but often the phase difference lies somewhere in between. The different cases are schematically illustrated in Figure 1.5. The phasing in Figure 1.5a characterizes the tide as a progressive wave, a situation often found along closed coastlines and continents. Figure 1.5b depicts a standing wave, where effectively a superposition occurs between an incident and reflected wave; this situation is sometimes found in tidal basins and estuaries.

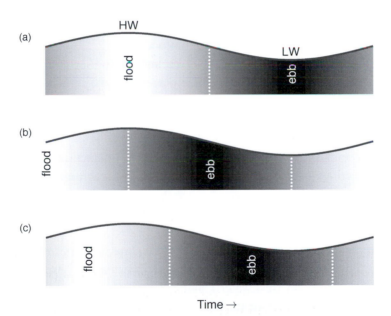

Figure 1.5 A sketch depicting three different scenarios concerning the phase relation between tidal currents and sea level at a given location. The time interval covers one tidal period, with a high water (HW) and a low water (LW). Beneath the surface, the current is depicted, with white and light gray indicating the flood phase, and dark gray to black, the ebb phase. Phases of slack water, i.e., the turnings of the tide, are indicated by vertical dashed lines. In (a), maximum flood coincides with high tide, and maximum ebb with low tide. In (b), slack waters coincide with high and low tides. In (c), we have an intermediate situation.

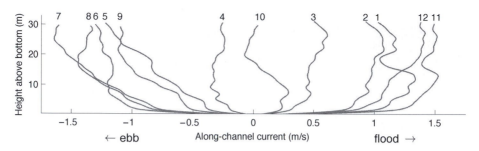

Figure 1.6 Vertical tidal profiles in an inlet, following a tidal cycle. Profiles are numbered according to their temporal sequence; the time interval between them is slightly over an hour, on average. These profiles were obtained from Acoustic Doppler Current Profile (ADCP) measurements at an anchor location, indicated by the black triangle in Figure 1.7. Based on Gerkema et al. (2014).

More often, however, one encounters an intermediate state in those regions, as in Figure 1.5c. Only in the case of Figure 1.5b does the phase of a dropping sea level coincide with ebb, and a rising sea level with flood.

The relation between tidal currents and sea level is further compounded when different periodicities are involved. At some places, one may find a daily signal in sea level variation, whereas ebb and flood occur twice a day – or vice versa.

To add yet another element of complexity to the tidal currents, we should note that the picture sketched so far – of water particles going simply "back and forth" – represents only a special case. In general, tidal currents involve both horizontal directions (west–east and south–north, say), each component with its own phase. In other words, at a moment when the current in one direction vanishes, the other may still be nonzero. In such cases, the notions of ebb, flood, and slack water become blurred.

In coastal regions, the *vertical structure* of tidal currents is an important feature (Figure 1.6). For example, in the tidal exchange in estuaries – with saline water entering during flood and freshwater leaving during ebb – most of the transport takes place in the upper part of the water column, where currents are strongest, which has consequences for the vertical layering of fresh and saline water during the different stages of the tidal cycle (a process called tidal straining).

In coastal regions, another concept closely related to the tidal current is often used, namely the *tidal prism*: the volume of water that flows in and out of a tidal lagoon or basin with flood and ebb. By way of example, values are mapped in Figure 1.7 for a barrier-island system that features a number of inlets. The largest tidal prisms occur in the deepest and broadest inlets.

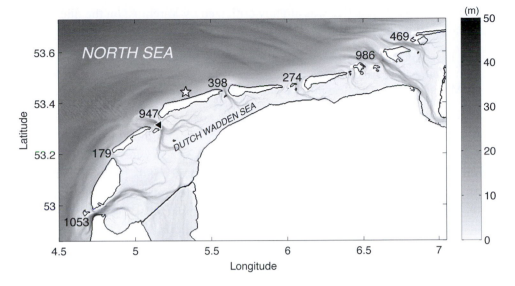

Figure 1.7 Bathymetric map of the western Wadden Sea (cf. Figure 1.4 for geographical orientation), with the tidal prisms indicated at the inlets, in 10^6 m^3; these values are taken from Gräwe et al. (2016). The gray scale shows water depth in meters. Also indicated are the locations of tide gauge *Terschelling Noordzee* (white star), and of *Vlie Inlet* (black triangle) where the current record from Figure 1.6 was obtained. A record from the tide gauge is examined in Section 1.4.

Exercise

1.1.1 At a certain location, a tracer is released during high tide. The tidal excursion is known to be 10 km. How far from the location of release does the tracer move during the following tidal period in the case of Figure 1.5a. And how far in the case of Figure 1.5b?

1.2 Generation and Dissipation of Tides

Semidiurnal and diurnal variations in sea level were already observed and documented in ancient times. In many coastal areas, they were plainly visible. Perceptive observers also noticed a connection with the phases of the Moon (higher tides shortly after full or new Moon). This is mentioned, for instance, by Pliny the Elder in his *Naturalis Historia*, where he also states that the cause of the tides lies in the Moon and Sun. Interestingly, the very idea that the Moon would have any influence on terrestrial affairs (including the tides) was later dismissed by some as a descent into occultism.[1] Others just furnished the cause with a name, as if this amounted to

---

[1] e.g., Isaac Vossius in his *De motu marium et ventorum* (1663).

an explanation. This practice was already criticized by Varenius; on the question of what causes the tides, he writes,[2]

> they have no other reply than to say that the Moon pulls the water along by a hidden form of attraction [*sympathiam*]. But these are mere words, which say no more than that the Moon produces the effect in some unknown way. That does not resolve the question.

In Newton's theory, this "hidden form of attraction" is called gravity. In itself, the word does not explain anything, of course; and Newton refrained from speculating about the deeper causes of gravity. The strength of his theory lies elsewhere: 1) the universal nature of the concept of gravity, which at once puts the Moon and Sun on an analogous footing as tide-generating bodies 2) the mathematical formulation of the force of gravity (the inverse-square law), which shifts the focus from the unanswerable *why* to the more productive *how*. In particular, this fundament allows us to derive the expression for the tide-generating force and to examine its implications. Here, we start with a simple qualitative argument that explains why there are predominantly two high and two low waters a day.

1.2.1 Cause of the Tides

We imagine the Earth to be entirely covered with a layer of water (Figure 1.8a). According to Newton's law of gravity, the Earth and Moon exert an attractive force on each other. The first point to note is that the layer of water would stay as it is if the force exerted by the Moon acted equally everywhere on Earth. In other words, *spatial variations* in this force are needed to reshape the layer. These variations are indeed there, since the strength and direction of the force vary; in particular, it becomes weaker at greater distance. Lunar gravity is more strongly felt at position A than at the center of the Earth C (or D and E), since the latter lies farther away from the Moon. The force thus tends to elevate A away from the center C. By the same token, the center C experiences a stronger pull than position B, so that the water at B tends to stay behind, as it were. The upshot is that there is a net force pulling water towards A and C, and away from B and D, as indicated by the arrows in Figure 1.8a.

How the water responds to this force is a different matter altogether. If we assume that the response of the water is immediate, in the sense that the layer always stays adjusted to the force, then it would take the shape depicted in Figure 1.8b. This is the hypothetical *equilibrium tide*. During the daily rotation of the Earth on its axis, one passes two bulges and therefore experiences two high waters.

However, the tides that we experience on Earth look very different; the bulges never really form because they continuously radiate tidal waves, producing a complex pattern of waves in the ocean basins. Moreover, the continents stand in

---

[2] Varenius in his *Geographia generalis* (1650).

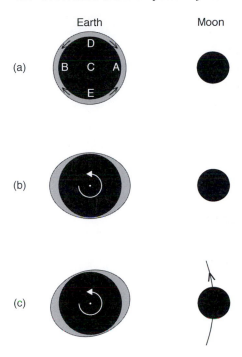

Figure 1.8 Sketches of the Earth and Moon, the former covered by a layer of water (in gray). In (a) we take a layer of uniform depth as a starting point. However, the tide-generating force exerted by the Moon (indicated by arrows) tends to reshape this layer. A hypothetical frictionless and instantaneous response of the water is indicated in (b), with high waters at A and B, and low waters water at D and E. The Earth is seen on top, i.e., the axis of the daily rotation sticks out perpendicularly from the paper, and the sense of rotation is indicated by the white arrow. In the course of a day, i.e., during a full cycle of rotation, one encounters two high waters and two low waters. In (c) the movement of the Moon in its orbit is indicated (note that its sense of revolution is the same as that of the Earth's spin on its axis). Tidal friction now retards the bulges, putting them out of line with the Earth–Moon axis.

their way. However, this does not change the fact that there are two high waters per day, for the tidal waves have the same periodicity as the forcing. The bulges are indeed more about the forcing than about the actual response of the water.

Meanwhile, the key to understanding the origin of tides has been identified: namely, the *spatial variation* in lunar gravity, rather than lunar gravity as such. The same reasoning applies to the tides generated by the Sun's gravitational force.

1.2.2 Moon and Sun

To pursue this last point a little further, we compare the Moon and the Sun with respect to gravity and its derivative. The force of gravity between Moon and Earth (F_m), and between Sun and Earth (F_s) is, respectively,

$$F_m = G \frac{M_e M_m}{r_m^2}, \qquad F_s = G \frac{M_e M_s}{r_s^2}, \qquad (1.1)$$

where G is the gravitational constant, M_e the mass of the Earth, and M_m (M_s) the mass of the Moon (Sun), and $r_{m,s}$ the mutual distance in each case. The values are listed in Table 3.1. We thus find for the ratio of the forces

$$\frac{F_m}{F_s} = \frac{M_m r_s^2}{M_s r_m^2} = 0.0056.$$

The gravitational pull exerted by the Sun is about 180 times stronger than that by the Moon.

Let us now consider the spatial variation of gravity, expressed as the derivative with respect to distance r. This changes the square of r in the denominators of (1.1) into a cubic power, giving more weight to the distance from the Earth, which works in favor of the nearby Moon. Indeed so much so that it reverses the situation:

$$\frac{dF_m/dr}{dF_s/dr} = \frac{M_m r_s^3}{M_s r_m^3} = 2.18.$$

When it comes to tidal generation, the Moon is more weighty than the Sun.

1.2.3 Energy Flows

For centuries, tidal energy has been harvested by humans. This was done already in the late Middle Ages by constructing tide mills that store water during rising tides, which is subsequently released during low tides to drive a wheel. More recently, tidal currents have been exploited in regions with macrotides (Figure 1.3) for the production of electricity. All this constitutes only a minute part of the total tidal energy that is dissipated in the oceans and seas.

Most of the tidal dissipation takes place by friction in the bottom boundary layer of the continental shelf seas. Tidal currents are typically two orders of magnitude stronger in those shallow seas than in the deep ocean. This difference is greatly amplified in the dissipation rate (i.e., friction force times current velocity), which goes with the cubic power of the current.

Before discussing a more detailed picture of tidal dissipation, we turn to the opposite side of the equation: where does the tidal energy come from in the first place? We start by revisiting the imaginary bulges of the equilibrium tide; see Figure 1.8c. Unlike in Figure 1.8b, there is now friction, causing a time lag in the response of the water. The bulges now occur shortly after passing the Earth–Moon axis, as the Earth went on with its daily rotation. This introduces an asymmetry in the setting that has consequences for the Earth as well as for the Moon. The Moon exerts a stronger pull on the bulge facing the Moon than on the opposite one. This

Figure 1.9 Left: astronaut Buzz Aldrin brings the study of tides a step forward during the Apollo 11 mission, July 21, 1969; in his right hand he holds the Laser Ranging Retro-Reflector (LR3). Middle: the LR3 left on the Moon. Right: experiment conducted at the Laser Ranging Facility, Goddard Space Flight Center. Photographs courtesy of NASA.

was also the case in Figure 1.8b, but the net force now has a component that is parallel to the Earth's surface, and thus exerts a torque, which acts against the sense of daily rotation. This causes a slowing of the Earth's spin. Meanwhile, the bulges themselves exert an asymmetric gravitational force on the Moon: in particular, the nearest bulge (again the most influential one) pulls the Moon forward in its orbit.

Thus far, the argument is freewheeling,[3] but the laws of motion control and connect the effects in a precise way. The upshot is that tidal friction causes three things to change: the Earth's spin on its axis (ω_e), the Moon's angular velocity in its orbit (ω_m) and the Moon's distance from the Earth (r). These three quantities are connected by two equations representing fundamental mechanical principles; details are provided in Box 1.1. Briefly, the first equation states that the total angular momentum of the Earth–Moon system is conserved. If the Moon's effect on the bulges is to retard the Earth's spin, decreasing its angular momentum, then the bulges do the opposite with the Moon: increasing the angular momentum of its orbital movement. Furthermore, the Moon's orbital movement is characterized by two parameters: its orbital angular velocity and the radius of its orbit. They may both change but always need to obey Kepler's third law, which serves as the second equation.

This results in two equations for the rate of change of ω_e, ω_m, and r. If one of them is measured, the other two can be calculated. The distance between the Earth and Moon can be accurately determined by laser ranging techniques. Put simply, the distance is calculated from the time it takes for light (from a laser-beam) to return after reflection from sophisticated mirrors deployed on the Moon

---

[3] In particular, the notion of "bulges" needs to be taken figuratively; they merely serve to illustrate the broader point that *any* break of symmetry in the enveloping water layer with respect to the Earth–Moon axis will create a torque and induce the dynamical effects outlined in Box 1.1.

(Figure 1.9). These experiments have demonstrated that the lunar distance, amidst its large periodic variations, increases at a net rate of 3.82±0.07 centimeters per year. From this value, the rate of loss of mechanical energy in the Earth–Moon system can be calculated (Box 1.1, Equation (1.7)), which is equivalent to the tidal dissipation. It is important to keep in mind that the actual causal chain works the other way round: it is the amount of tidal dissipation that dictates the change in the Earth's spin and the lunar orbital parameters.

The corresponding increase in the length of day is 2.3 milliseconds per century. This looks like a small amount, but the increments add up quadratically with time; as a result, the cumulative effect already amounts to a lapse of several hours over the course of a few millennia (Exercise 1.2.1). Empirical evidence supports the notion of a lengthening of the day through geological and historical times, though mostly at a slower rate than at present. Sources range from pre-Cambrian sedimentary tidal rhythmites, counts of daily growth rings per year in Devonian coral fossils (as the days were shorter, more fitted into a year) to reported solar eclipses on Babylonian clay tablets. In ancient times, astronomers carefully documented where and when total solar eclipses occurred. Modern astronomical calculations can reconstruct these events, but assuming a constant length of day would place them westward of the actual location (Figure 1.10). From this discrepancy, the change in the length of day can be inferred. Here, another factor plays a role as well, for changes in the mass distribution on Earth (e.g., post-glacial rebounds) modify the moment of inertia of the Earth, which is compensated by a change in the rotation rate to conserve angular momentum. As a consequence, changes in the length of day cannot be as unequivocally ascribed to tidal dissipation as the recession of the Moon.

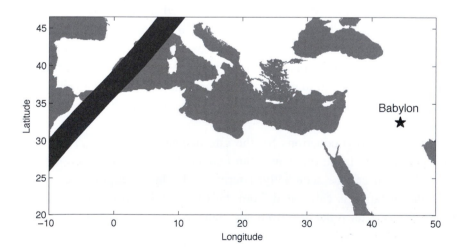

Figure 1.10 A total solar eclipse was observed in Babylon on 15 April 136 BC. Calculating backward with the current length of day, the eclipse would instead have occurred westward, as indicated by the black diagonal band. After Stephenson (2003).

Overall, the available evidence makes it plausible that the present rate of tidal dissipation is higher than during most of the Earth's past. For one thing, the Moon would have orbited improbably (and cataclysmically) close to Earth in the relatively recent geological past (i.e., 1.5 billion years ago) if we calculate backward with the present recession rate of the Moon, contrary to evidence that the Moon is much older. Global tide models run for past configurations of the continents indeed show a lower dissipation rate. This subject of *paleotides* concerns not only different configurations of the continents but also sea level variations. During the glacial cycles of the Pleistocene, for instance, sea level varied by more than 100 meters, which strongly marks the areal extent of the continental shelf seas and hence the tidal characteristics and dissipation.

Returning to the present rate of tidal dissipation, we sketch the pathways from sources to sinks in Figure 1.11. The most certain number is the one in the upper left corner, the input of energy from the Earth–Moon system, which can be calculated from (1.7), using the measured recession rate of the Moon. By comparison, the contribution from solar tides (upper right corner) is relatively small. The tidal energy is split between the atmosphere, the solid Earth and the oceans. The first two are very minor compared to the ocean. Incidentally, atmospheric tides are more significant than this number suggests, because there is an additional source of tidal energy (not included in this diagram): the daily cycle of thermal heating. They are called

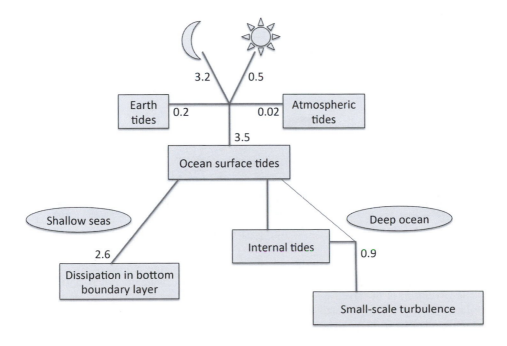

Figure 1.11 Estimates of the pathways of tidal energy towards dissipation. Values are in Tera Watt (TW, Tera=10^{12}) and taken from Munk and Wunsch (1997).

thermal tides to distinguish them from those generated by lunar or solar gravity. Tides in the solid Earth (Earth tides) are briefly discussed in Box 1.2.

In the ocean, tidal dissipation follows essentially two pathways. One is the dissipation by bottom friction, mainly in the shallow continental shelf seas, which we already mentioned. The other route involves a transfer of energy from surface tides to internal tides. They are waves that propagate in the interior of the ocean, making the subsurface levels of equal density (*isopycnals*) oscillate while the surface itself hardly moves by comparison. They arise when the currents of surface tides meet a bottom slope: the periodic movement up and down the slope brings the isopycnals into motion, from which internal waves radiate at the tidal frequency, the *internal tides*. They spread into the abyssal oceans where they can break near bottom slopes (Figure 1.12) or by instabilities due to strong shear currents, finally leading to small-scale turbulent mixing. This source of mixing has oceanographic implications that extend well beyond the topic of tides itself: a certain amount of mixing is needed to maintain the large-scale meridional overturning circulation. Present estimates are that the tides provide about half of the required energy, while the other half comes from wind input.

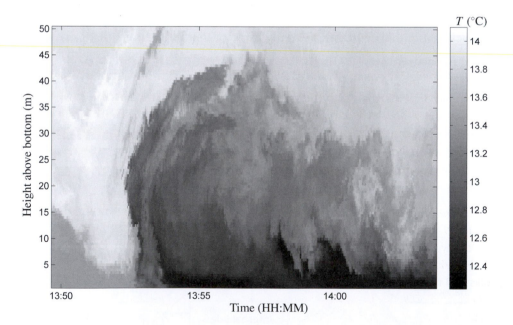

Figure 1.12 An example of a large backward-breaking internal wave over the slope of Great Meteor Seamount, Atlantic Ocean; local water depth is 550 m. The evolution is observed by means of 101 fast-sampling temperature sensors, placed at a mooring at equal spacing between 0.5 and 50 m above the bottom. Figure adapted from Van Haren and Gostiaux (2012), with permission from The Oceanography Society.

Exercises

1.2.1 The present increase in the length of day is 2.3 milliseconds per century (i.e., the contribution of tidal dissipation).

a) Show that the cumulative time lapse is $42 \times n_c^2$ seconds, where n_c is the number of centuries.

b) Figure 1.10 illustrates the hypothetical location of a solar eclipse if the length of day had been constant at its present value (black diagonal); this is westward of the actual location (Babylon). Why *west*ward?

1.2.2 The slowing of the Earth's spin due to tidal dissipation actually releases more energy than goes into tidal dissipation, for a certain amount of energy is gained by the lunar orbital motion at the expense of the Earth's spin.

a) Calculate the relative proportion.

b) Show that, despite the Moon's gain in energy, both its angular and orbital velocity *decrease*. How is this paradox resolved?

Box 1.1 Tidal Dissipation, Length of Day and Lunar Orbit

To present the problem in its simplest form, we make a number of assumptions:

- We ignore the Sun and consider only the Earth–Moon system;
- Instead of looking at the orbital movements of the Earth and Moon around their common center of mass (see Figure 2.1), we assume that the Moon figures as a satellite, with a relatively small mass, in the Earth's gravitational field, ignoring the Earth's own orbital movement;
- We assume that the Moon traces a circular orbit in the equatorial plane.

Let the mass of the Earth (Moon) be M_e (M_m) and their mutual distance, from center to center, r. The Earth spins on its axis at angular velocity ω_e; the orbital angular velocity of the Moon is ω_m.

We adopt expressions for angular momentum and kinetic energy known from classical mechanics. The angular momentum of the Moon's orbital movement is $M_m \omega_m r^2$. For the Earth's spin, angular momentum can be expressed as $I_e \omega_e$, where I_e is the moment of inertia. (Note that for a spherical body of radius a and with uniform mass distribution the moment of inertia would be $\frac{2}{5} M_e a^2$, but the Earth has a relatively dense core, which makes the moment of inertia smaller: $I_e \approx \frac{1}{3} M_e a^2$.) Conservation of total angular momentum for the Earth–Moon system can thus be expressed as:

$$M_m \omega_m r^2 + I_e \omega_e = \text{constant}. \tag{1.2}$$

To calculate the rate of change, we take the time derivative (indicated by a dot), which gives

$$M_m\dot{\omega}_m r^2 + 2M_m\omega_m r\dot{r} + I_e\dot{\omega}_e = 0. \tag{1.3}$$

This equation features the changes in the length of day (via $\dot{\omega}_e$), and in the length of month (via $\dot{\omega}_m$), and the recession of the Moon ($\dot{r}$).

Kepler's third law implies that the Moon's orbital movement must obey

$$\omega_m^2 r^3 = GM_e. \tag{1.4}$$

Hence

$$2r\dot{\omega}_m + 3\omega_m\dot{r} = 0. \tag{1.5}$$

Using (1.3) and (1.5), we express $\dot{\omega}_m$ and $\dot{\omega}_e$ in terms of $\dot{r}$:

$$\dot{\omega}_m = -\frac{3\omega_m}{2r}\dot{r}, \qquad \dot{\omega}_e = -\frac{M_m\omega_m r}{2I_e}\dot{r}. \tag{1.6}$$

The total mechanical energy (i.e., kinetic plus potential) of the system is given by

$$E = \frac{1}{2}I_e\omega_e^2 + \frac{1}{2}M_m\omega_m^2 r^2 - G\frac{M_e M_m}{r} = \frac{1}{2}I_e\omega_e^2 - \frac{1}{2}M_m\omega_m^2 r^2,$$

where we used Kepler's third law (1.4) to simplify the expression. Finally, using (1.6), we can express the rate of change in the energy as

$$\dot{E} = \frac{1}{2}M_m\omega_m(\omega_m - \omega_e)r\dot{r} \tag{1.7}$$

This represents the loss in mechanical energy induced by tidal dissipation. At present, a day is shorter than a month (i.e., in terms of frequencies: $\omega_e > \omega_m$), but the tidal dissipation comes to a halt once they have become of equal length, a phenomenon called *tidal locking*. This has happened to the Moon, which is why we always see the same side of the Moon.

1.3 Nontidal Variations and Mean Sea Level

In the next section, we examine a year-long record from a tide gauge to identify various characteristic features of the tidal signal. First, we briefly consider other nontidal variations of sea level to sketch the dynamic backdrop against which tides occur.

We start with wind-generated gravity waves, whose periods range from several seconds to a few tens of seconds; during stormy conditions, they can attain amplitudes of several meters or more. The locally generated waves (called sea) occupy the high frequency range, while the slower ones (swell) are the remainder of wind waves that were generated elsewhere, sometimes as far as thousands of kilometers away. The period band of these waves is quite distinct from the semidiurnal and diurnal tidal periods, which makes it easy to filter out their effect if we want to isolate the tidal signal from a sea level record. For this, we need to take averages over a period that is long enough to cancel out the contribution of the wind waves, but short enough to be well below the tidal time scales. In practice, a period of about ten minutes suits this purpose.

Apart from generating waves, the wind also produces surges or depressions of sea level, especially during storms. They occur over large areas (tens of kilometers or more) and have typical time scales of several hours (with a short but vehement storm) or days (with sustained strong winds from a certain direction). Storm surges are particularly pronounced in semi-enclosed basins in coastal regions, due to funneling effects, where they can reach heights of several meters. This happens for example in the southern North Sea (Figure 1.4) during northwesterly storms. Conversely, during strong winds from the opposite direction, the sea level becomes anomalously low. Because of their timescales, which are comparable to those of tides, the wind-generated surges and depressions cannot be simply removed from a sea level record by bandpass filtering without also affecting the tidal signal. Moreover, the signals interact and cannot be regarded as fully independent. A storm surge affects the water depth and hence the phase speed of tidal wave propagation. Conversely, surges propagate differently during high than during low tides.

There is yet another meteorological factor that affects the sea level: atmospheric pressure. Spatially, the troposphere is organized in patterns of moving high and low pressure areas. The sea surface is pushed down where atmospheric pressure is high and, conversely, is lifted up where atmospheric pressure is low. This is known as the *inverted-barometer effect*. The effect amounts to sea level variations up to several decimeters. For example, in the southern North Sea, at a weather station near the location indicated by the triangle in Figure 1.7, longterm records show that atmospheric pressure typically lies between 985 and 1040 mbar in this region. Theoretically, this corresponds to a vertical span of sea level of about half a meter, since 1 mbar in atmospheric pressure is approximately equivalent to the hydrostatic pressure of a column of water of 1 cm height. In practice, the effect is not always noticeable because the response of sea level takes time to build up; moreover, the signal is often contaminated by concurrent wind surges or depressions.

The combined effect of wind surges (or depressions) and atmospheric pressure primarily causes variations in sea level on a time scale of several hours to a few days. However, the occurrence and intensity of these phenomena varies from year to year. As a result, annual mean values of local sea level can vary by a few decimeters between different years while also showing decadal variability. On still longer time scales, climatic trends can be detected in sea level change, typically a rise at a rate of a few millimeters per year at present, but with considerable regional variability.

From this short inventory we can draw two conclusions that are relevant for the analysis of tides. First, sea level records generally represent an aggregate of tides and wind surges or depressions, as well as variations due to the inverted-barometer effect, which act on similar time scales. In contrast, wind-generated gravity waves can easily be filtered out from the record because they have a much shorter time scale. Second, the notion of "mean sea level" is not evident, for the outcome depends on the time scale over which one takes the mean. The ideal picture of the tide as an oscillation with respect to a well-defined mean level thus faces an obstacle in reality. In practice, therefore, high and low tides are stated with respect to a fixed vertical reference (datum), defined regionally, which usually falls somewhere within the range of typical values for annual mean sea level. Incidentally, in nautical maps it is common to use other vertical references to indicate the water depth, for example mean low tide or lowest low tide. It is important to be aware of which reference is being used!

1.4 Example of a Sea Level Record

Sea level is routinely recorded at many harbors and other coastal locations world-wide. We show an example in Figure 1.13, a one-year record from a tide gauge in the southern North Sea, at the location marked by an asterisk in Figure 1.7. From this example we infer a number of tidal characteristics.

By visual inspection, we discern recurrent variations on several time scales as well as occasional irregularities. To begin with the latter, the signal exhibits finger-prints of the wind. In box A, water levels reach the lowest values of the year. This was a period with persistent southeasterly winds; the water was blown away from the coast and drained off from the barrier-island basins into the North Sea, resulting in anomalously low water levels in the coastal area. The opposite occurred around late October, when a northwesterly storm created a brief but large surge, which clearly stands out in box B. These are just two of the most conspicuous examples, but the wind interferes through much of the year, especially in autumn and winter.

We now turn our attention to the tides. The dominant signal is unmistakably semidiurnal: throughout the record, we observe two high and two low waters in a day. The record thus confirms the global map shown in Figure 1.2, which indicates that tides have a predominantly semidiurnal character on the west European Shelf. Counting the peaks over the whole year, we find 706 low waters and 705 high waters. This means that the average semidiurnal tidal period is 12 hours and 25 minutes.

However, the record also reveals a number of additional variations. By way of example, we select a month (box C) to illustrate some of them. Here, the semidiurnal oscillation is seen to undergo two kinds of modulation. At the

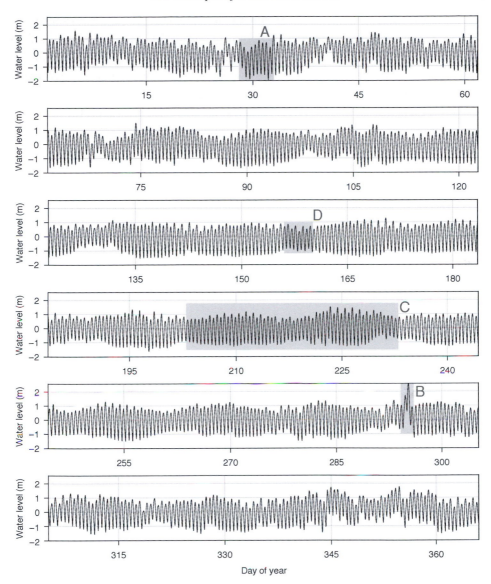

Figure 1.13 Sea level record from a tide gauge in the North Sea coastal area, at the position shown by an asterisk in Figure 1.7. The data covers the full year 2014, at 10-minute intervals. The vertical reference is a datum used in the Netherlands (NAP). Data supplied by the Dutch governmental agency Rijkswaterstaat.

beginning, middle, and end of the record in box C, the tidal range is relatively small, whereas at both intervals in between, it attains maximum values. The same pattern is seen elsewhere in the record, especially during spring and summer (but less clearly during parts of autumn and winter, when the signal becomes fuzzier due to wind effects). This cycle of higher and lower tidal ranges is called the *spring-neap cycle*: tidal ranges are large during spring tides and small during neap tides. The spring-neap cycle is generally a conspicuous feature in tidal records around the world. A larger tidal range (during spring tides) works in two ways: high waters are higher than normal, and low waters lower than normal. For this reason, the spring-neap cycle is of great significance for navigation and coastal ecology. In coastal areas, places that are normally inundated may fall dry at low waters during spring tides. In addition, the spring-neap cycle affects the tidal currents. Stronger currents during spring tides increase the potential for erosion, resuspension, and transport of sediment. In short, the spring-neap cycle is a notable factor in the coastal environment. The record also faintly hints at a modulation of the spring-neap cycle itself: they appear to have an alternating strength. During the second cycle in box C, tidal ranges reach slightly larger values than during the first cycle. This alternation extends to the cycles directly before and after box C as well.

A second kind of modulation shows up in box C in the high waters. Most of the time, they have alternating heights: successive high waters are unequal. This is called the *diurnal inequality*. It means that there is one higher high water and one lower high water in a day. However, the diurnal inequality is not present all the time: it comes and goes approximately two times in box C. In this example, the absence of the diurnal inequality coincides more or less with neap tides, but this is an accidental simultaneity. The contrary is the case in box D, for instance, where the diurnal inequality is clearly present during neap tides. The spring-neap cycle and the cycle of the diurnal inequality both occur roughly twice a month, but with slightly different periodicities, so that they are out of sync.

We can explore the high and low waters further by plotting them in isolation, leaving out the rest of the record. Thus, in Figure 1.14, we see the sequences of high waters (in black) and low waters (in gray). The diurnal inequality now stands out more clearly. It occurs in the high waters but not noticeably in the low waters. In Figure 1.14 we have added one extra piece of information, namely the distinction between daytime and nighttime. This reveals that the moments of higher and lower high waters change through the year: in the first and last few months of the year, higher high waters generally occur during nighttime (closed circles), but in the intermediate period, during daytime (open circles).

In summary, we have qualitatively identified a number of tidal cycles. Anticipating more detailed explanations in later chapters, we already briefly sketch their astronomical background:

- The dominant signal in this record is the semidiurnal oscillation, at a mean period of 12 hours and 25 minutes. There are not quite two tidal periods in a day as long

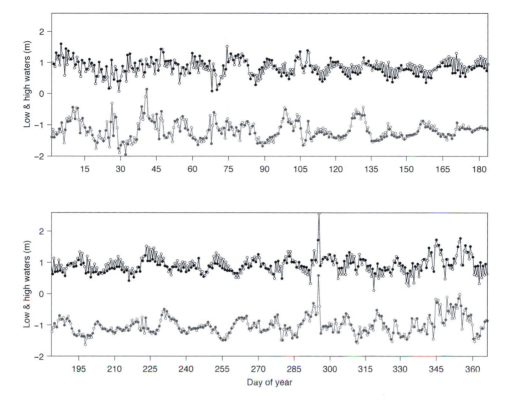

Figure 1.14 Record of high waters (in black) and low waters (in gray), extracted from Figure 1.13. The timing of the high and low waters is also indicated, as daytime (6a.m.–6p.m., open circles) and nighttime (6a.m.–6p.m., closed circles). Data supplied by the Dutch governmental agency Rijkswaterstaat.

as we refer to our usual day of 24 hours, but the two periods fit very neatly in an alternative "day" of 24 hours and 50 minutes. This length of time has a special astronomical significance: it is the interval between two successive transits of the Moon across the meridian.

- The record shows a bimonthly (i.e., twice a month) alternation of higher and lower tidal ranges, the spring-neap cycle. The origin of the spring-neap cycle lies in the periodic alignment of the Moon, Earth, and Sun, which happens every 14.8 days, at full and new Moon. Around these lunar phases, spring tides occur, because the tide-generating forces due to the Sun and Moon add up constructively.
- An alternation occurs in successive spring tides: maximum tidal ranges alternate slightly. This inequality waxes and wanes at a period of 206 days. During one such period, spring tides are largest around full Moon, during the next period, around new Moon. This cycle is related to the fact that the Moon traces an elliptic orbit around the Earth, combined with a slow turning of the ellipse itself.

- High waters show a diurnal inequality, which comes and goes at a period of 13.7 days. The origin of the diurnal inequality lies in the declination of the Moon, i.e., the fact that the orbit of the Moon has an angle with the equatorial plane. Every 13.7 days the Moon stands at a maximum angle, alternately in the northern and southern hemispheres.
- A yearly cycle occurs in the daytime versus nighttime occurrence of the higher high waters in the diurnal inequality. This phenomenon finds its origin in the inclinations of the lunar orbital plane and the ecliptic with respect to the equatorial plane.

1.5 Scope and Outline of this Book

In this introductory textbook we treat tides in *oceans, coastal seas, and basins*. The basic principle of the tide-generating force applies more generally, but the response to this force depends on the properties of the medium in question, including the nature of its boundaries, which sets the oceans and seas apart from the solid Earth and atmosphere. Tides in the solid Earth (Earth tides) are briefly discussed in Box 1.2. Tides in the atmosphere (mostly thermal in origin, as noted in Section 1.2.3) imprint an oscillation on the ocean surface via atmospheric pressure. For the main diurnal solar component, this imprint is even more important than the part generated directly in the ocean itself; for the main semidiurnal solar component, the imprint offers a small additional signal.

Beyond the Earth, tides are a common phenomenon in planet-moon systems. We are used to looking at tides on Earth due to the Moon's gravity, but for the giant planets the most spectacular tides appear on the moons rather than on the planets. In some cases, tidal heating is thought to be responsible for volcanic activity and strong thermal convection in the interior, e.g., for Saturn's moon Enceladus. As a matter of fact, even in our own Moon weak half-monthly tides are present (due to the ellipticity of its orbit around the Earth), which have been linked to the occurrence of moonquakes.

Expressions for the tide-generating force and corresponding potential are derived in Chapter 2. They contain a number of astronomical parameters, such as the distance from the Moon (or Sun) and the angle of the tide-generating body with the equatorial plane, the declination. Both vary periodically in time at cycles of (half-)months and longer periods – or in the case of the Sun, (half-)years.

Before moving on with an analysis of the tide-generating potential, we provide in Chapter 3 a descriptive overview of the relevant periodicities in the Earth–Moon–Sun system, which are crucial to our understanding of the tidal signal in the ocean and the origin of the multitude of periodicities it contains.

The forcing is precisely known at any given moment in time if one knows the positions of the Moon and Sun relative to the Earth. In principle, one can then insert the forcing term in the equations of motion, known from geophysical fluid dynamics, and calculate the response – the tides. This approach is followed in numerical global tide models, but the complexities (due to the presence of continents and more generally the propagation of forced waves on a sphere) preclude analytical treatment. While we do not dwell on the direct connection between forcing and response in this book, much can still be learned from analyzing the tide-generating potential and the propagation of tidal waves separately.

Chapter 4 focuses entirely on the tide-generating potential, which contains a multitude of periodicities stemming from the orbital motions. The terms in the potential can be expanded in a sum of sinusoids with related periodicities. Knowing these frequencies is extremely useful, for the same frequencies can be expected to appear in the response. This idea lies at the basis of the harmonic method of tidal prediction. The tidal signal at any place (e.g., a harbor) is conceived as a superposition of sinusoids with these frequencies, which are known a priori. The amplitude and phase of the sinusoids can be estimated from past local records of sea level. Once this has been accomplished, the local tidal signal can be predicted by evaluating the sum of sinusoids for future moments in time.

The dynamic approach in Chapter 5 is to examine free-wave solutions (i.e., without considering the forcing: the waves are just assumed to be there) at tidal frequencies for various simple configurations. This reveals the spatial pattern of tides, conceived as a wave phenomenon. These models, which can be treated analytically, are highly idealized but nonetheless explain key aspects of the tidal waves as we observe them in the oceans. In particular, the central role of continental slopes in guiding tidal waves is elucidated.

Tides in continental shelf and coastal seas are primarily a co-oscillation with the tides propagating in the adjacent ocean. Yet they form a special category with characteristics that set them apart from ocean tides (Chapter 6). Due to the shallowness of the system, tidal currents are generally much stronger than in the deep ocean. They play a key role in erosion and transport of sediment as well as transport of dissolved or suspended matter. The strong currents also imply a larger role of nonlinear processes (via advection and friction). This produces an additional class of tidal frequencies.

On the one hand we have tidal frequencies that originate from the myriad of periodicities involved in the motions of the Moon and Earth with respect to each other and with respect to the Sun. We call them *astronomical constituents*. They can be deduced by looking at the heavens; not a single measurement in the ocean is needed to deduce those tidal frequencies. For the additional class of tidal frequencies in shallow seas, the *shallow-water constituents*, it is the other way round: no astronomical measurement can offer a clue of their existence. Instead, we find them in data from tide gauges and in current velocity measurements. To understand them theoretically, we have to consider the hydrodynamic laws governing the water motions, whose nonlinear terms create their own multitude of tidal frequencies out of the astronomical ones.

Finally, internal tides, mentioned already briefly in Section 1.2.3, is a wide-ranging subject. In Chapter 7, we offer a concise introductory tour through models that describe their basic properties.

Box 1.2 **Earth Tides, Load Tides and Self-Attraction**

The tidal forcing sketched in Figure 1.8a acts not only on the ocean but also on the solid Earth, creating *Earth tides*. Their dynamics is very different, and in fact much more straightforward to model, than for ocean tides. This is because tidal waves in the solid Earth are fast, in the sense that they traverse the Earth many times within a tidal period. They are thus able to adapt in a quasi-immediate way to the relatively slowly changing forcing: they essentially follow the tide-generating potential, with certain factors (called Love numbers) that account for the elastic properties of the solid Earth. Earth tides result in periodic radial displacements of the Earth, which have amplitudes of up to 20 cm. The ocean participates in this movement.

A terrestrial observer relates tidal elevations in coastal seas to local water depth, as measured at a tide gauge which is fixed to the solid Earth. Earth tides then pass unnoticed, for they make the sea surface and bottom move up and down in unison. However, if observed by satellite altimetry, the movement of the ocean surface is measured with respect to the Earth's center. Here, a correction for Earth tides needs to be made lest the corresponding sea surface movement is interpreted as an ocean tide proper.

Global ocean tides manifest themselves as large-scale wave patterns. The solid Earth beneath them experiences the high and low tides as increases and decreases in pressure. Its response amounts to a vertical movement of up to 5 cm in amplitude, which is broadly 180° out of phase with the ocean tide (i.e., the seafloor moves down under high tides, and up under low tides); these oscillations are called *load tides*. However, the response is in fact not merely local, but stretches out spatially, even over land. Finally, another factor comes into play as *self-attraction*, since the redistribution of mass in high and low tides and corresponding changes in the gravity field affect the height of the water column. Since the redistribution of mass involves the ocean as well as the solid Earth, self-attraction is included in models together with load tides (SAL: self-attraction and loading). They have an effect on the signal measured in bottom pressure sensors and at tide gauges.

Further Reading

In this chapter, we touched upon several topics, but for some of them, a more detailed discussion would be beyond the scope of this textbook. Here, as in later chapters, we provide a guide to further reading.

For a review on Earth tides, load tides, and self-attraction, see Baker (1984); global maps of load tides, for the main semidiurnal and diurnal constituents, are shown in Pugh and Woodworth (2014). A formalism for the inclusion of self-attraction and loading in numerical tidal models is detailed by Ray (1998). Agnew (2007) presents an overview on the theory of Earth tides. An estimate of the dissipation involved in Earth tides was derived by Ray et al. (2001). For atmospheric tides, the book by Chapman and Lindzen (1970) remains the classic reference. The imprint of atmospheric diurnal and semidiurnal tides on the ocean was examined by Ray and Egbert (2004) and Arbic (2005), respectively.

For tides and tidal heating on moons, see for example Spencer (2011) on Saturn's moon Enceladus. For our own Moon, the connection between tides and moonquakes was examined by Toksöz et al. (1977).

Evidence on the change in the length of day on geological and historical time scales is wide-ranging. Deposits of marine sediments in tidal environments are modulated by the tidal cycles. In these sedimentary cyclic rhythmites, the layering may reflect how many days are contained in a month, and how many months in a year. Williams (2000) reviews the evidence from Precambrian deposits of changes in the length of day. In the Phanerozoic Eon, the availability of fossils with growth rings can offer clues, as reviewed by Hansen (1982). In historical times, records of solar eclipses by Babylonian, Chinese, and Arab astronomers can be used to estimate the more recent changes in the length of day, as explained by Stephenson (2003).

On the subject of paleotides, model results on tidal dissipation in the geological past are presented by Bills and Ray (1999) and Green et al. (2017). On a time scale of about 400–450 million years, Earth goes through cycles of the formation and subsequent breaking-up of a supercontinent; model results by Green et al. (2018) suggest that tidal dissipation is weakest during the phase of a supercontinent and that we are presently close to maximum tidal generation and dissipation. In earlier configurations of the continents, some seas were nearly isolated. An example is the epicontinental sea in the present-day northwestern Europe during the Carboniferous, for which tides were modeled by Wells et al. (2005). Effects of sea level change on tides, through glacial cycles, are examined by Egbert et al. (2004) and Arbic et al. (2008). Tamisiea and Mitrovica (2011) and Pugh and Woodworth (2014) discuss present-day sea level change and its regional variability.

Tidal dissipation, via internal tides, and its role in ocean mixing is reviewed by Munk and Wunsch (1997, 1998) and by Garrett and St. Laurent (2002). On geological time scales, changes in tidal dissipation may play a role in climatic variations, as discussed by Munk and Bills (2007).

In Section 1.2, we briefly mentioned a few historical elements regarding the cause of the tides. Ekman (1993) presents a short overview on the history of tides; for a comprehensive treatise, see Cartwright (1999).

The general surveys on ocean tides by Doodson (1958) and Cartwright (1978), under identical titles, are still enlightening. From a historical point of view, they marked transitional moments: the authors summarized the knowledge at that point but also signalled a certain impasse, as further progress on the understanding of ocean tides awaited new technical developments, "high-speed machines" and satellite altimetry, respectively.

Popular books on tides, at the level of this introductory chapter, include the classic work by Darwin (1911) and the more recent textbook by Open University (1989).

2

Tidal Forcing

In this chapter we derive an expression for the tide-generating force and the corresponding potential. We take the Moon as the tide-generating body, but exactly the same reasoning applies to the Sun. In both cases, we can make an approximation that exploits the fact that the tide-generating body is situated at a large distance from the Earth, compared to the Earth's radius.

2.1 Forces in the Orbital Motions of the Earth–Moon System

In the previous chapter, we simplified the movements in the Earth–Moon system by assuming that the Moon circles a motionless Earth (Figure 1.8c), except for the daily rotation on its axis. This attributes to the Earth a special place that in fact it does not have: the Earth and Moon both trace orbits, in which they are kept by each other's gravitational force. Their orbits are different, but they go around the same virtual, motionless point: their common center of mass (see Box 2.1).

Three phases of their monthly orbital movements are sketched in Figure 2.1a. The center of mass C lies closer to the Earth than to the Moon because the mass of the Earth is larger. As a result, the Earth's monthly orbit is smaller. In reality, the center of mass lies even within the Earth, but for illustrative purposes it is drawn outside of it, which is immaterial to the argument that follows. The same holds for another simplification we made by assuming the orbits to be circular, whereas they are actually weakly elliptic (cf. Figure 3.5).

In Figure 2.1b, we take a closer look at the precise nature of the Earth's monthly orbital movement. We have marked an arbitrary but fixed geographical location on Earth (open circle), besides its center (black circle). They can be followed in their orbital movement, again depicted in three snapshots. The crucial point is that their orientation remains the same throughout; the white circle is always facing upward in the figure. In other words, the orbital movement, in itself, is a pure *translation*. (In reality, of course, the Earth also turns on its own axis, but this daily rotational

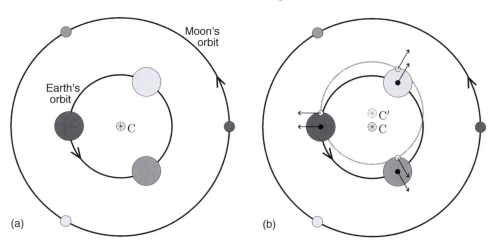

Figure 2.1 (a) The Earth and Moon circling around their common center of mass (C), making a full orbit in the course of a month; snapshots are shown in three shades of gray. (b) A replication, with additional details on the Earth's orbital movement. Two fixed terrestrial positions are indicated: the center (black circle) and a peripheral point (open circle). Notice that the Earth revolves but does not rotate: its *orientation* remains unchanged through this purely orbital movement. The peripheral point describes a similar circle as the center of the Earth, but slightly shifted, orbiting C′ instead of C. In every phase of the orbital movement, the two points occupy identical places in their respective circles; hence, the centrifugal forces (indicated by the outward-pointing arrows) must be the same for both positions at any time.

movement is independent of its monthly orbital movement and we here consider the latter in isolation.) The peripheral position traces a circle (depicted in dashed gray), which is a shifted version of the Earth's orbit; the latter actually represents the orbit of the center of the Earth. The different positions trace their respective circles in unison; for example, when one of them occupies the outer left, the other does, too. This implies that the centrifugal force associated with their circular movements is always identical, as indicated by the outward-pointing arrows. Finally, instead of taking the position of the open circle in Figure 2.1b, we may take any other position on or within Earth, with the same results.

The upshot is that at any moment, all points on Earth experience the *same* centrifugal force, both in strength and direction, as depicted by the outward arrows (in light gray) in Figure 2.2. The gravitational force exerted by the Moon, on the other hand, differs in strength and direction for different points on Earth (black arrows). It becomes weaker at larger distances and is always directed to the center of the Moon, which implies a varying angle with the Earth–Moon axis. The combined effect of the centrifugal force and lunar gravity is indicated by the dark gray arrows in Figure 2.2.

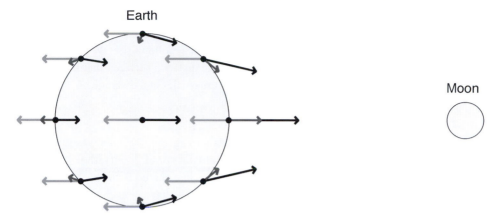

Figure 2.2 Snapshot of the monthly orbital movement of Earth and Moon. The forces are indicated by arrows: the centrifugal force (light gray), the gravitational lunar force (black), and the resultant tide-generating force (dark gray), obtained by adding up the two vectors. For illustrative purposes, the Moon is placed very close to the Earth to render spatial differences in lunar gravity more clearly visible.

The only point that experiences no net force is the center of the Earth, where the centrifugal force is exactly balanced by lunar gravity,[1] creating an apparent state of weightlessness. An analogous situation occurs for a space station circling the Earth. In such a station, astronauts experience no gravity, even though they are kept in their orbit by the Earth's gravity field. This paradoxical state arises because the gravitational force is exactly balanced by the outward centrifugal force associated with the station's circular orbit. In the present case, Earth is the "space station," circling the center of mass of the Earth–Moon system, and kept in its orbit by lunar gravity. But, again, the balance of forces holds only at the center of the Earth.

At positions closer to the Moon, a weight is experienced as a net pull toward the Moon, whereas at positions farther from the Moon, the opposite happens and objects tend to float away, as indicated by the net forces in Figure 2.2. In reality, of course, nothing gravitates toward the Moon or floats away from it, because Earth's own gravity force is strong enough to hold everything together. However, what it cannot prevent is that water starts to flow *along* the Earth's surface in response to the imbalance of forces.

---

[1] If we consider an elliptic orbit instead of a circular one, the outward-pointing arrows no longer represent a purely centrifugal force, but also accelerations in the radial direction. However, this has no consequences for the derivation of the tide-generating force, because the magnitude of the outward force is simply deduced from the fact that it is in balance with lunar gravity at the center of the Earth (which remains true for an elliptic orbit), which saves us from the need to actually attribute a specific meaning to the outward-pointing force.

Box 2.1 **Center of Mass**

The distance from the center of mass can be calculated as follows. Let M be the mass of one body, m the mass of the other, and let r be the distance between them, measured from center to center. Then the center of mass lies at a distance of

$$\frac{m}{M+m}r \qquad (2.1)$$

from the center of the first body. This distance becomes smaller with decreasing m/M.

For two bodies orbiting each other under the mutual force of gravity, the closed orbits take the form of ellipses; the center of mass lies at a shared focus of the ellipses. In the special case of a circular orbit, it simply lies at the center of each circle (as in Figure 2.1a).

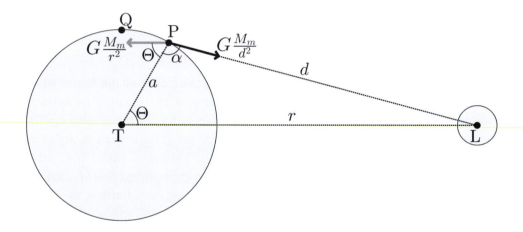

Figure 2.3 The centrifugal force (gray arrow) and lunar gravity (black arrow) acting on terrestrial position P. The center of the Earth is denoted by T; the center of the Moon, by L. The distance between these centers is r. The distance between P and the center of the Moon is d; a is the radius of the Earth. Angles Θ and α are used in the decomposition of the forces.

2.1.1 Derivation of the Tractive Force

We derive an expression for the tide-generating force at an arbitrary position P, as depicted in Figure 2.3. We formulate the problem in terms of forces per unit of mass, but follow common practice and still refer to them as "forces," although strictly speaking they are accelerations.

The magnitude of the centrifugal force can immediately be deduced from the fact that it balances lunar gravity at the center of the Earth, which lies at distance r from the Moon. Hence the centrifugal force is GM_m/r^2, where M_m denotes the mass of the Moon and G is the gravitational constant.

We define angles $\Theta = \angle LTP$ and $\alpha = \angle TPL$. Notice that Θ does not represent latitude, even though Figure 2.3 might give that impression. (For a more complete sketch, see Figure 2.7, where latitude is introduced as ϕ and is clearly contrasted with Θ.) At position P, the centrifugal force (gray arrow) and lunar gravity (black arrow) can each be decomposed into a component normal to the Earth's surface and a tangential one.

The normal components, being of the order of GM_m/r^2, are about 3×10^5 times smaller than Earth's own gravity (see the values listed in Table 3.1). They merely create a minute change in the weight of a parcel of water. For the tangential components, on the other hand, there is no such counteracting force, so they will be the most significant ones. For the moment, we therefore focus on the tangential components. With regard to their sign, a convenient reference is the sense of Θ, which we choose positive in the counterclockwise direction.

For the centrifugal force, the tangential component thus reads

$$F_{c,t} = G \frac{M_m}{r^2} \sin \Theta. \tag{2.2}$$

This component is positive for positions P in the first and second quadrants (pointing counterclockwise), and negative in the third and fourth quadrants (pointing clockwise).

The angle between lunar gravity and the tangential at P is $\alpha - \pi/2$, so its tangential component can be written as

$$\begin{aligned} F_{g,t} &= -G \frac{M_m}{d^2} \cos(\alpha - \pi/2) \\ &= -G \frac{M_m}{d^2} \frac{r}{d} \sin \Theta, \end{aligned} \tag{2.3}$$

where we applied the sine rule with respect to the triangle TPL in Figure 2.3:

$$\frac{\sin \alpha}{r} = \frac{\sin \Theta}{d}.$$

By including a minus sign in (2.3), we have accounted for the fact that the tangential component of lunar gravity is negative in the first and second quadrants, and positive in the third and fourth quadrants. Finally, by summing the contributions (2.2) and (2.3), we obtain the tangential component of the tide-generating force, also known as the *tractive force*,

$$F_t = F_{c,t} + F_{g,t} = \frac{GM_m}{r^2} \left(1 - \frac{r^3}{d^3} \right) \sin \Theta. \tag{2.4}$$

With the cosine rule

$$d^2 = r^2 - 2ra \cos \Theta + a^2$$

(again with reference to Figure 2.3), we can express d in terms of a, r and angle Θ. Hence (2.4) becomes

$$F_t = \frac{GM_m}{r^2}\left(1 - \frac{r^3}{[r^2 - 2ra\cos\Theta + a^2]^{3/2}}\right)\sin\Theta$$

$$= \frac{GM_m}{r^2}\left(1 - [1 - 2(a/r)\cos\Theta + (a/r)^2]^{-3/2}\right)\sin\Theta. \qquad (2.5)$$

Now, the ratio a/r, the Earth's radius over the distance to the Moon, is small: about $1/60$ (see Table 3.1). To exploit this fact, we apply the binomial series (A.9) to the terms in square brackets in (2.5). This results in

$$\boxed{F_t = -\frac{3}{2}\frac{a}{r}\frac{GM_m}{r^2}\sin 2\Theta + \cdots.} \qquad (2.6)$$

Here we also used the trigonometric identity (A.8). The dots stand for quadratic terms, and higher, in a/r; we neglect them henceforth.

The resulting tractive force (2.6) is inversely proportional to the *cube* of r, the Earth–Moon distance, as already anticipated in Section 1.2.2. The strength and direction of the force vary with Θ, the angle of the radial TP with the Earth–Moon axis. The force F_t is negative for positions P in the first and third quadrants, where it is directed clockwise, and positive in the second and fourth quadrants, where it is directed counterclockwise. In Figure 2.3 we took an arbitrary cross-sectional plane of the Earth as a starting point for the derivation. The same result is obtained for any other cross-sectional plane containing the Earth–Moon axis. In other words, the tractive force is cylindrically symmetric with respect to the Earth–Moon axis, as illustrated in a spherical representation of the Earth in Figure 2.4. This holds not only for (2.6), but for higher-order terms as well.

In Figure 2.4, the half-sphere facing the Moon is the exact mirror image of the opposite half-sphere. This symmetry characterizes the lowest-order expression (2.6), but is not carried over to higher orders. This is already evident from the original representation of the forces in Figure 2.2, which exhibits an asymmetry in the net force (dark gray) between the half-spheres. However, in that figure the asymmetry is greatly exaggerated by placing the Moon close to the Earth (to render the net effect of the centrifugal force and lunar gravity more clearly visible). The asymmetry diminishes as the Moon is placed farther away. Recall that the approximation (2.6) is indeed based on the assumption of small a/r ratio, i.e., a distant Moon.

Exercise

2.1.1 Extend the development of (2.5) up to order $(a/r)^2$ and show that the expression for the tractive force then becomes:

$$F_t = -\frac{3}{2}\frac{a}{r}\frac{GM_m}{r^2}\left[\sin 2\Theta + \frac{a}{r}(5\cos^2\Theta - 1)\sin\Theta\right].$$

Show that the symmetry between the half-spheres, i.e., the one facing the Moon and the opposite one (cf. Figure 2.4), is now broken.

Earth

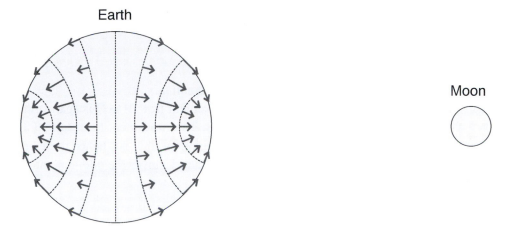

Moon

Figure 2.4 The tractive force (2.6) depicted for a spherical Earth. Notice the cylindrical symmetry with respect to the Earth–Moon axis.

2.1.2 Tide-Generating Potential

The potential of the tide-generating force at P represents the amount of work done by the force when we move a parcel from P to the center of the Earth, where the force is zero. This result is independent of the path, so we may take a mathematically convenient one.

We first move the parcel along the Earth's surface from P to Q (Figure 2.3). The path increment ds is $a\,d\Theta$, with boundaries $\Theta_P = \Theta$ and $\Theta_Q = \pi/2$, so the integral of the tractive force in (2.6) becomes

$$
\begin{aligned}
V_1 &= \int_P^Q ds\, F_t = a \int_{\Theta_P}^{\Theta_Q} d\Theta\, F_t \\
&= -\frac{3}{2}\left(\frac{a}{r}\right)^2 \frac{GM_m}{r} \int_{\Theta_P}^{\Theta_Q} d\Theta\, \sin 2\Theta \\
&= -\frac{3}{2}\left(\frac{a}{r}\right)^2 \frac{GM_m}{r} \cos^2\Theta,
\end{aligned}
\tag{2.7}
$$

where we used the trigonometric identity (A.6).

The second track is from Q right to the Earth's center T. This path is always normal to the centrifugal force (see Figure 2.2), so the only contribution to the work comes from lunar gravity, which must be the difference between the lunar gravitational potentials at Q and T, i.e.,

$$
V_2 = -\frac{GM_m}{(r^2 + a^2)^{1/2}} + \frac{GM_m}{r}
$$

$$= \frac{GM_m}{r} \left(1 - [1 + (a/r)^2]^{-1/2} \right)$$

$$= \frac{1}{2} \left(\frac{a}{r} \right)^2 \frac{GM_m}{r} + \cdots , \tag{2.8}$$

where we used again the binomial series (A.9) for small a/r.

A check of the signs is in order. For V_1, the movement from P to Q is against the tractive force, hence the work in (2.7) is negative. The movement from Q to T is with the force of lunar gravity (its along-path component, that is), hence the work in (2.8) is positive. The sum of the contributions V_1 and V_2 gives the tide-generating potential at P:

$$V = -\frac{1}{2} \left(\frac{a}{r} \right)^2 \frac{GM_m}{r} (3 \cos^2 \Theta - 1). \tag{2.9}$$

This expression serves as the basis for further examination in the following sections and in Chapter 4. We note that in the literature there is no uniformity with regard to the sign on the right-hand side of (2.9), as it depends on whether one takes the strictly physical potential (as we do here) or its negative counterpart.

It can be easily checked that taking the gradient of the potential in the tangential direction leads us back to the tractive force (2.6), i.e.,

$$F_t = -\frac{1}{a} \frac{\partial V}{\partial \Theta}. \tag{2.10}$$

Through V_2, the potential (2.9) also includes the effect of the normal (i.e., radial) component of the tide-generating force, F_n. Interpreting radius a as a coordinate rather than as a constant, we can extract this component from the potential by

$$F_n = -\frac{\partial V}{\partial a}. \tag{2.11}$$

Of special significance are the equipotential contours (i.e., lines on which the potential is constant), which are normal to the tide-generating force. Again, we here interpret radius a as one of the independent variables, the other variable being angle Θ. It is convenient to change to Cartesian coordinates, with the Earth's center at the origin: $x = a \cos \Theta$ and $y = a \sin \Theta$. With this transformation, we find from (2.9) that the equipotential contours are given by

$$2x^2 - y^2 = \text{constant},$$

which represent hyperbolas; they are shown in Figure 2.5a. The asymptotes $y = \pm x \sqrt{2}$ describe the zero potential lines, which correspond to $\cos^2 \Theta = 1/3$ in (2.9). Particles subject to the tide-generating force move from high to low values of the potential, crossing the contours perpendicularly. Thus, they tend to move toward the x-axis, the Earth–Moon axis, and away from the Earth. Although the

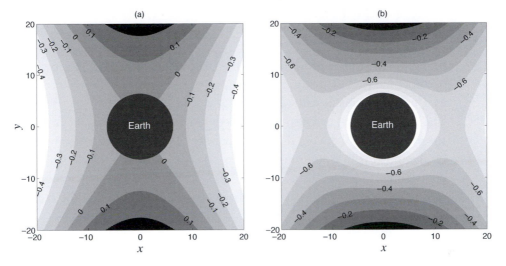

Figure 2.5 Equipotential contours. In (a) for the tide-generating potential alone; in (b) together with terrestrial gravity. To render the departure from the concentric contours of gravity visible in (b), the mass of the Moon is taken much larger (× one million). The sphere at the center represents the Earth. The Moon (not shown) is located on the x-axis, far outside of the range of the figure.

Earth is depicted in Figure 2.5a, it stands idle, as it were, for the graph represents the sole effect of the tide-generating force; at this stage, the potential due to terrestrial gravity is ignored. Adding the latter reshapes the equipotential contours completely; they become visually indistinguishable from the concentric circles that belong to terrestrial gravity. To illustrate qualitatively the effect of the tide-generating potential in the presence of terrestrial gravity, we artificially amplify the mass of the Moon. The result is shown in Figure 2.5b. Close to the Earth, the concentric circles of terrestrial gravity are reshaped into ellipse-like contours. They remind us of the "bulges" of Section 1.2.1, the hypothetical equilibrium tide. Far away from the Earth,[2] terrestrial gravity loses importance relative to the tide-generating force, and the contours start to resemble those of Figure 2.5a.

Exercises

2.1.2 a) Show that the normal component of the tide-generating force, in the setting of Figure 2.3, is given by

$$F_n = G \frac{M_m}{d^3} (r \cos \Theta - a) - G \frac{M_m}{r^2} \cos \Theta .$$

---

[2] Notice that we should stay within the range $a \ll r$, lest the assumption underlying (2.9) is violated.

Hint: the cosines of Θ and α satisfy the simple trigonometric identity $a = r\cos\Theta + d\cos\alpha$.

   b) Follow a similar procedure as in (2.4) to (2.6) to derive the lowest-order approximation of F_n in terms of a/r.

   c) Check your result with (2.11).

2.1.3 The potential (2.9) can also be obtained without first deriving the tractive force. Calculate the work done by lunar gravity and the centrifugal force on the direct path from P to T (Figure 2.3) and show that their combined contributions lead to (2.9).

2.1.4 Following up the result from Exercise 2.1.1, demonstrate that the corresponding potential is

$$V = -\frac{1}{2}\left(\frac{a}{r}\right)^2 \frac{GM_m}{r}\left[(3\cos^2\Theta - 1) + \frac{a}{r}(5\cos^3\Theta - 3\cos\Theta)\right].$$

This higher-order correction to (2.9) offers a relatively small but still significant contribution to lunar tidal components, but can be neglected for solar tides (for which a/r is much smaller).

2.2 Declination

Thus far in this chapter, we have treated the Earth as a sphere with no orientation whatever. This changes with the introduction of the polar axis, associated with the Earth's daily rotation, as illustrated in Figure 2.6. Importantly, the equatorial plane has an angle with the lunar orbital plane, called the *inclination*. This affects both the daily and monthly appearance of the tides.

We already drew attention to the fact that the tractive force on the half-sphere facing the Moon is the exact mirror image of the opposite half-sphere (Figure 2.4). However, this symmetry is not carried over to a daily symmetry for a fixed position on Earth, precisely because of the axial tilt. For example, at geographical location F in Figure 2.6 we have a different tractive force than half a day later, when it is situated at F'. Herein lies the origin of the diurnal inequality that we discerned in a tide gauge record discussed in Section 1.4.

The angle of the Moon's actual *position* with the equatorial plane is called the *declination*. For a constant angle of the lunar orbital plane with the equator (the inclination), the declination goes through varying angles as the Moon traces its monthly orbit: from high over the northern hemisphere (as in Figure 2.6), to right above the equator (zero declination), to low over the southern hemisphere, etc. In terms of the absolute value of the declination, this amounts to a cycle at a

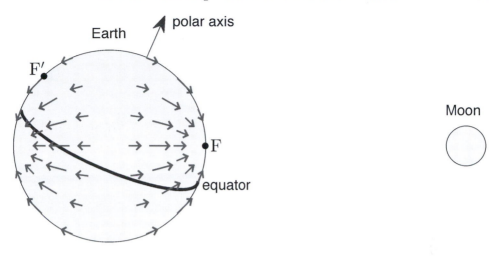

Figure 2.6 Replication of Figure 2.4, but now with the additional features of the polar axis and the equatorial plane.

half-monthly period. The diurnal inequality, which owes its existence to the declination, reflects this cycle and comes and goes twice a month.

2.3 Tide-Generating Potential in Terrestrial Coordinates

In Section 2.1.2 we derived the tide-generating potential in terms of angle Θ. The resulting expression (2.9) is elegant but impractical, because we do not normally refer to positions on Earth by indicating the angle with the Earth–Moon axis.

For this reason, we now adopt a geographical perspective, referring angles to longitude and latitude. All relevant angles are indicated in Figure 2.7. Again, we have an arbitrary geographical location P whose radial TP has an angle Θ with the Earth–Moon axis TL, but the Moon now comes off the page; it stands right above position Z. Through P and Z we draw the meridians that meet the equator at P' and Z', respectively. The longitudinal angle between those meridians is Ψ. Position P lies at latitude ϕ (i.e., $\angle$PTP') and the Moon is situated at latitude δ (i.e., $\angle$LTZ'), which is the declination. Relative to a given position P, at latitude ϕ, the location of the Moon in the sky is specified by the angles Ψ and δ. We adopt an identity known from spherical trigonometry to express Θ in terms of those angles:

$$\cos \Theta = \sin \phi \sin \delta + \cos \phi \cos \delta \cos \Psi . \tag{2.12}$$

For a derivation of this identity, see Box 2.2.

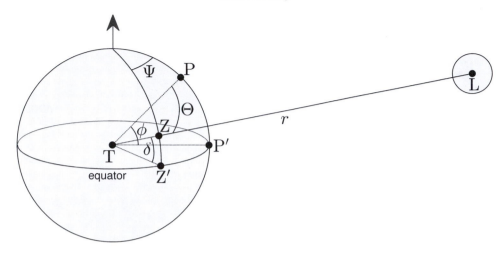

Figure 2.7 Angles defining the position of the Moon and the geographical location P in terms of longitudinal and latitudinal angles. Notice that the Moon stands right above point Z and comes off the page.

Box 2.2 **Trigonometric Identity**

Here we outline a proof of identity (2.12), following Proudman (1953).

We refer to the angles and positions defined in Figure 2.7. Consider a unit vector pointing from the Earth's center T to Z. This vector can be decomposed into a component $\cos\delta$ lying in the equatorial plane, along TZ', and a component pointing toward the North Pole, $\sin\delta$. We project each of these components onto the line TP. For the latter, this yields the contribution $\sin\phi\sin\delta$. The former can first be projected into the plane TPP' (i.e., onto the line TP'), which gives $\cos\delta\cos\Psi$, since Ψ is the angle $\angle Z'TP'$. Finally, the projection of this vector onto TP is $\cos\phi\cos\delta\cos\Psi$. Adding up the two contributions, we find that the projection of the original unit vector onto TP is

$$\sin\phi\sin\delta + \cos\phi\cos\delta\cos\Psi.$$

But this projection can also be done directly via the angle Θ, which gives $\cos\Theta$, so we must have

$$\cos\Theta = \sin\phi\sin\delta + \cos\phi\cos\delta\cos\Psi,$$

which is the identity (2.12).

We substitute (2.12) in (2.9) and organize the terms according to the occurrence of angle Ψ. For the moment, we leave aside the coefficient in (2.9) and focus on the factor $3\cos^2\Theta - 1$. After substitution of (2.12), this factor becomes

$$3\sin^2\phi\sin^2\delta + \tfrac{3}{2}\sin 2\phi\sin 2\delta\cos\Psi + 3\cos^2\phi\cos^2\delta\cos^2\Psi - 1,$$

where we used the trigonometric identity (A.8) for ϕ and δ to simplify the second term. In the third term, we use (A.6) to obtain

$$3\sin^2\phi\sin^2\delta + \overbrace{\tfrac{3}{2}\sin 2\phi\sin 2\delta\cos\Psi}^{\sim\cos\Psi} + \overbrace{\tfrac{3}{2}\cos^2\phi\cos^2\delta\cos 2\Psi}^{\sim\cos 2\Psi} + \tfrac{3}{2}\cos^2\phi\cos^2\delta - 1.$$

The terms without Ψ can be grouped together; after some manipulation, this results in

$$\tfrac{9}{2}(\sin^2\phi - \tfrac{1}{3})(\sin^2\delta - \tfrac{1}{3}).$$

Hence, the tide-generating potential (2.9) can be written

$$V = -\frac{3}{4}\left(\frac{a}{r}\right)^2\frac{GM_m}{r}\left[\overbrace{3(\sin^2\phi - \tfrac{1}{3})(\sin^2\delta - \tfrac{1}{3})}^{\text{constant}} + \overbrace{\sin 2\phi\sin 2\delta\cos\Psi}^{\sim\cos\Psi}\right.$$
$$\left. + \overbrace{\cos^2\phi\cos^2\delta\cos 2\Psi}^{\sim\cos 2\Psi}\right]. \tag{2.13}$$

With reference to the longitudinal angle Ψ, we find three kinds of terms: a constant term, one that oscillates with Ψ, and one that oscillates at the double frequency 2Ψ.

Until now, we have treated the setting of Figure 2.7 as a static one, frozen in time as it were. However, it is easy to see how time comes into play. First of all, the Earth spins on its axis. This means that the longitudinal angle Ψ makes a full cycle in one "day," something like $\Psi = \Omega t$, where Ω is the corresponding angular frequency and t denotes time. The relevant time period is here linked to the orientation with respect to the Moon, which itself moves as well, so we will have to look carefully exactly what we mean by a "day" (see Chapter 3). Regardless, we can now see how the tides make their appearance: the $\cos\Psi$ term in (2.13) corresponds to a diurnal (i.e., daily) oscillation, while $\cos 2\Psi$ corresponds to a semidiurnal (i.e., twice a day) oscillation. With that, we have identified in (2.13) the essential part of the tidal forcing. The constant term in (2.13) implies that *mean* sea level is also modified by the tide-generating force.

From the coefficient of $\cos\Psi$ it is clear that the diurnal tides owe their existence to the declination δ, confirming the qualitative argument in Section 2.2. If the Moon moved in the equatorial plane, the declination would always be zero and the term with $\cos\Psi$ would vanish altogether.[3]

---

[3] Recall that we restrict ourselves here to the lowest-order term in the tide-generating potential; at higher orders, (weak) diurnal tides appear even without declination (Exercise 2.3.2).

Time enters the potential not only via Ψ, but also via declination δ and lunar distance r. The Moon has a monthly period in declination as it goes from maximum values of δ (standing high in the northern or southern hemispheres) through vanishing δ as it crosses the equatorial plane. The distance r varies over a monthly period, too, as the Moon traces an elliptic orbit. These variations imply that the "constant" first term in (2.13) actually varies in time, but at relatively long periods (bimonthly, monthly). The tides resulting from this forcing are known as *long-period tides*. Parameters δ and r also appear in the factors of the diurnal and semidiurnal terms and hence produce slow *modulations* of these tides. This is further explored in Chapter 4, but first we have a closer look at the main periodicities involved in the motions of the Earth–Moon–Sun system, in Chapter 3.

Exercises

2.3.1 In a cataclysmic event, the Moon is hit by a large comet. The Moon is kicked into a different orbit and now traverses a meridional plane; once a month the Moon passes over the North and South poles. Discuss qualitatively the changes in the tidal characteristics.

2.3.2 Following up the result from Exercise 2.1.4, show that at this order there is a diurnal tidal component even in the absence of declination.

2.4 Combined Tractive Forces of Moon and Sun

In previous sections, we have derived the lunar tractive force and corresponding tide-generating potential. Exactly the same reasoning applies to the Sun; we only have to replace some of the parameters. Instead of mass M_m we take the mass of the Sun, M_s. Distance r is now the distance between Earth and Sun, and declination δ refers to the position of the Sun with respect to the equatorial plane, which makes a yearly cycle.

The tractive forces induced by the Moon and Sun can be superposed to get their combined effect. Qualitatively, we can readily see that their superposition is not always acting in the same way: sometimes the tractive forces work against each other, sometimes they reinforce each other, depending on their alignment with the Earth. This is sketched in Figure 2.8, where we show the tractive forces at four different phases of the lunar cycle. During a monthly period, the Moon goes through the phases of new Moon, first quarter, full Moon, and third quarter. The tractive force is always directed toward the line connecting the Earth and the tide-generating body (Figure 2.4). Thus, when the Sun, Earth, and Moon are in line, the tractive

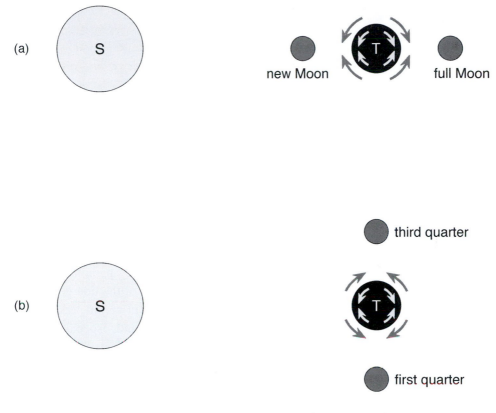

Figure 2.8 The Sun (S), Earth (T), and the Moon at four phases of the lunar cycle. The tractive force by the Moon is shown in gray arrows; the tractive force by the Sun in white arrows. In (a), the Moon, Earth, and Sun are aligned (new Moon and full Moon); the tractive forces work in the same direction, resulting in higher tides (spring tides). In (b), the Moon, Earth, and Sun are in quadrature (first and third quarters); the tractive forces now oppose each other, resulting in weaker tides (neap tides).

forces of the Sun and Moon work in the same direction (Figure 2.8a), creating spring tides. When they are in quadrature (Figure 2.8b), they work against each other, creating neap tides. Alignments and quadratures each occur twice a month, hence the period of the spring-neap cycle.

3

Celestial Motions

For the purpose of studying tides on Earth, we need to consider only two tide-generating bodies: the Moon and Sun. The influence of the other planets is negligible. In the previous chapter, we have identified two key parameters in the tide-generating potential (2.13): distance r and declination δ of the tide-generating body. This applies to the Moon and Sun separately, giving four parameters in total. They vary in time due to variations in the orbits of the Earth and, especially, of the Moon. In this chapter, we give a synopsis of the principal variations that affect the tides.

3.1 Sun-Earth System

3.1.1 Orbit

The Earth[1] traces an ellipse around the Sun, as sketched in Figure 3.1. Strictly speaking, it is the center of mass of the Earth–Sun system that lies at the focal point (see Box 2.1 for an explanation of the center of mass). For the Earth–Sun system, we can easily verify from the values listed in Table 3.1 that the center of mass lies very deeply within the Sun. So, for all intents and purposes, we can identify the focal point with the Sun itself. Basic properties of the ellipse, including its eccentricity (ε), are recapitulated in Box 3.1. The current eccentricity of the Earth's orbit is 0.017.

The point of closest distance to the Sun is called *perihelion*. In our epoch, this point is passed around the 3rd of January. The opposite point, at largest distance to the Sun, is called *aphelion*. These extremal points are also known as the *apsides*. According to Kepler's second law, a planet sweeps out equal areas in equal times in its movement around the Sun. This means that the orbital speed of the planet must

---

[1] We ignore the Moon for the moment, for its presence does not significantly alter this story.

Table 3.1 *Masses, radii, and distances in the Sun–Earth–Moon system. The third column refers to the semimajor axis, which in the case of the Moon varies considerably; the value stated here is an average.*

	Mass $(10^{24}$ kg)	Mean radius $(10^3$ km)	Distance from Earth $(10^3$ km)
Sun	1.989×10^6	6.955×10^2	1.496×10^5
Earth	5.9726	6.371	–
Moon	0.07342	1.737	3.844×10^2

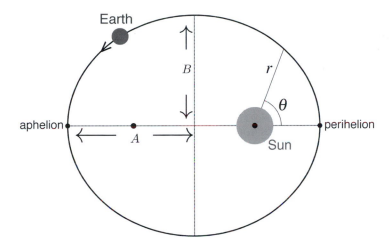

Figure 3.1 The orbit of the Earth, tracing an ellipse, with the Sun at one of its foci (which are marked by the two dots on the major axis). The semimajor axis (length A) and semiminor axis (B) are indicated. The polar coordinates r and θ are used in Box 3.1. For visual clarity, the eccentricity of the ellipse is exaggerated (here: $\varepsilon = 0.5$).

be anomalously high near perihelion, and anomalously low near aphelion. The time between successive passages of perihelion is called the *anomalistic year*, which has a length of 365.2596 days. This is not exactly the same as making a full orbit in the sense of being aligned to the same fixed star again, as measured by the *sidereal year* (from Latin *sidus* – star), which has a slightly shorter length of 365.2564 days. This is because the ellipse itself turns slowly in its orbital plane – the so-called perihelion shift or apsidal precession – and hence it takes an extra while for the Earth to catch up with perihelion. The perihelion shift amounts to 11.5" (i.e., arc seconds) per year, so that it takes more than a hundred thousand years for the ellipse to make a full turn. The eccentricity also varies, on a similar timescale. Both effects play a

Box 3.1 **Ellipses and the Kepler Problem**

Kepler discovered from Brahe's accurate measurements that the planets trace elliptical orbits around the Sun.

Geometrically, for every point on an ellipse, the sum of the distances to the two focal points is constant; this is how an ellipse is defined. The closer the focal points are to each other, the more the ellipse resembles a circle. The line through the focal points is called the major axis; the one perpendicular to it, the minor axis. The *eccentricity* ε of an ellipse is defined as

$$\varepsilon = \left(1 - \frac{B^2}{A^2}\right)^{1/2},$$

where A is the length of the semimajor axis, and B of the semiminor axis (Figure 3.1). The point on the orbit closest to the Sun, *perihelion*, lies at a distance $(1 - \varepsilon)A$; the most distant point, *aphelion*, at $(1 + \varepsilon)A$.

It is convenient to describe the ellipse in polar coordinates r and θ, with the origin at one of the focal points: r is the distance from the focal point and θ traces the angle along the orbit (Figure 3.1). The ellipse is then described by

$$r = \frac{A(1 - \varepsilon^2)}{1 + \varepsilon \cos \theta}$$

(an arbitrary constant angle may be added to θ to turn the ellipse).

Conservation of angular momentum means that

$$r^2 \frac{d\theta}{dt} = \text{constant},$$

which implies Kepler's second law, stating that equal areas are swept out in equal times. Thus, at perihelion the orbital speed is higher than at aphelion.

To find the position on the elliptical orbit as a function of time t is known as the *Kepler problem*. This problem is usually stated in terms of an auxiliary variable E (known in astronomical parlance as the eccentric anomaly), which appears in the following transcendental equation:

$$\omega t = E - \varepsilon \sin E \tag{3.1}$$

($\omega = 2\pi/T$, with T the orbital period). Once $E(t)$ has been obtained from (3.1), the orbital coordinates are found by substitution in the expressions

$$r = A(1 - \varepsilon \cos E), \qquad \cos \theta = \frac{\cos E - \varepsilon}{1 - \varepsilon \cos E}. \tag{3.2}$$

Different methods have been developed to solve (3.1). For example, E can be expressed in terms of Bessel functions J_n:

$$E = \omega t + 2 \sum_{n=1}^{\infty} n^{-1} J_n(n\varepsilon) \sin(n\omega t). \tag{3.3}$$

Alternatively, for weak eccentricity $\varepsilon \ll 1$, the second term on the right-hand side of (3.1) can be regarded as relatively small. Thus, we can solve (3.1) iteratively as $E_n = \omega t + \varepsilon \sin E_{n-1}$:

$$E_1 = \omega t, \qquad E_2 = \omega t + \varepsilon \sin \omega t, \quad \text{etc.} \qquad (3.4)$$

With this iterative process, one quickly gains accuracy as $\varepsilon \ll 1$.

significant role in long-term climatic variations (the Milankovitch cycles), but for our purposes, we can ignore these slow variations and regard the Earth's orbit as essentially static.

3.1.2 Obliquity

The Earth makes a daily rotation on its axis. This polar axis has an angle of 23.4° with the Earth's orbital plane, the *ecliptic* (which, from a terrestrial perspective, is the plane in which the Sun moves).

This obliquity is why we have different seasons (Figure 3.2). There are two moments in the year when the Sun crosses the equatorial plane; they are called the *equinoxes*. The meaning of this word is "equal nights": at the equinoxes, the length of the night is the same everywhere on Earth, 12 hours. There is a vernal equinox around March 20, marking the beginning of spring, and an autumnal equinox around September 22, marking the beginning of autumn. Two other special

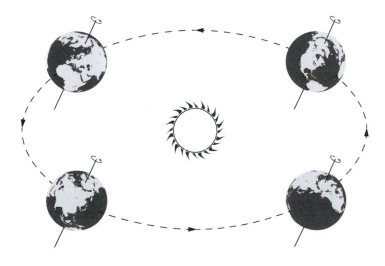

Figure 3.2 Obliquity of the Earth's axis with respect to the ecliptic, causing the four seasons. With reference to the northern hemisphere, starting on the lower left: summer, autumn, winter, and spring. Courtesy of Thomas Meutgeert.

days are the ones when the Sun reaches its highest point above the horizon in the northern or southern hemispheres, the summer and winter solstices, respectively, which fall around June 21 and December 21. They mark the beginning of summer and winter. As we have seen in the previous section, the Earth's orbital speed is higher near perihelion, and lower near aphelion. This has a noticeable effect on the length of the seasons; in our epoch, the winter half-year (defined as the period from autumnal to vernal equinox) is about a week shorter than the summer half-year (from vernal to autumnal equinox).

The *tropical year*[2] is defined as the time interval between two successive vernal equinoxes; its length is 365.2422 days. The tropical year is a convenient measure of time since it is, by definition, phase-locked to the seasons. For tides, too, it is an important period, for it represents the *solar declinational cycle*: with maximum declination at solstices and zero declination during equinoxes. The tide-generating potential (2.13), with now the Sun as the tide-generating body, shows that the forcing of solar diurnal tides is maximum during solstices, while solar semidiurnal tides are at their minimum; during equinoxes, it is the other way round.

The tropical year is about 20 minutes shorter than the anomalistic year (or the nearly identical sidereal year). This is due to the fact that the Earth's axis makes a slow westward precession, against the orbital movement – the so-called precession of the equinoxes (a full cycle takes about 26,000 years). Thus, the orbital position at which the vernal equinox occurs, is shifted slightly backward year after year. As a result, it takes less time to go from one vernal equinox to the next than to make a full orbital cycle.

As a point of reference, we may take a fixed star (the *sidereal* perspective), or the imaginary point of vernal equinox (the *tropical* perspective). They are not identical because the point of vernal equinox drifts slowly with respect to the stars.

3.1.3 Daily Rotation

The unit of time of a *day* is related to the Earth's rotation on its axis. A day can be defined in different ways, depending on the point of reference. The *solar day* is the time between two successive transits of the Sun across the meridian, as depicted in Figure 3.3 (central panels). Likewise, the *true sidereal day* refers to successive transits of a certain star (left panels), and the *lunar day* to successive transits of the Moon (right panels).

The difference between the various kinds of days arises from the orbital movements of the Earth and Moon, as sketched in Figure 3.4. We start at an arbitrary

---

[2] The adjective "tropical" comes from the Greek word for *turning* and refers to the turning of the Sun's meridian transit from its highest or lowest point over the horizon (at solstices).

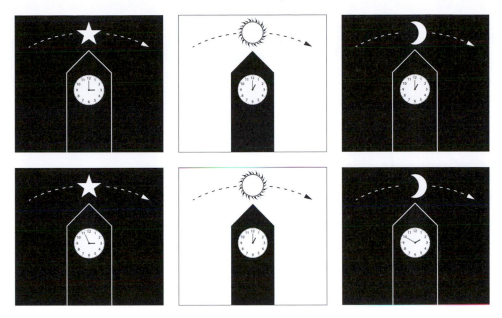

Figure 3.3 Moments of transit across the meridian of a star (left), the Sun (middle), and the Moon (right). Upper panels show the timing on a certain day; lower panels, the timing on the next day. The time interval defines the true sidereal day, solar day, and lunar day, respectively. The church tower stands south (north) for an observer in the northern (southern) hemisphere. Courtesy of Thomas Meutgeert. Inspired by a figure in Guérin (2004).

moment ("today"), at which the Moon, Earth, and Sun, as well as a fixed geographical position P, are conveniently aligned. The next day ("tomorrow"), the alignments are not occurring at the same time anymore.[3] The Earth makes one full rotation on its axis in 23 h 56 min 4.1 s; after this period, position P (now indicated by P′) is again facing the same distant star. We call this period the *true sidereal day*. In the meantime, the Earth has moved on in its orbit; as a consequence, it takes a little longer before P faces the Sun again (P″). This defines the familiar *solar day* of 24 h. Since the Moon progresses in its orbit as well, it takes still longer before P is again aligned to the Moon (P‴), which defines the *lunar day* of 24 h 50 min 28.3 s.

Figure 3.4 also suggests how the different days are related. For example, given the lengths of the solar day (D_{sol}) and the sidereal year (Y_{sid}), we can calculate the length of the true sidereal day (D_{sid}). In a true sidereal day, the Earth makes a full rotation over angle 2π. In a solar day, it makes an extra angle γ. Hence, we have

---

[3] To make the figure not overly complicated, we have frozen the position of the Earth during the stages from P′ to P‴, but in reality there is a concurrent movement along its orbit.

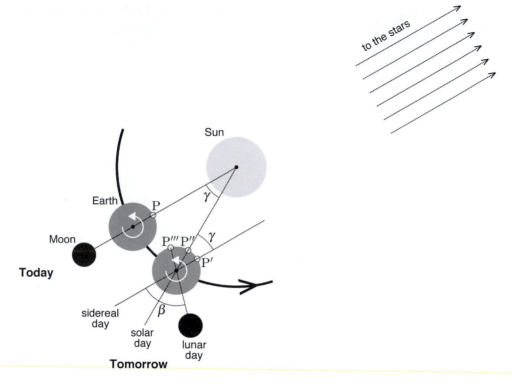

Figure 3.4 The difference between the true sidereal, solar, and lunar days, explained by the orbital motions of the Earth and Moon. Geographical position P is depicted a day later on three different occasions: aligned to a star (P′), the Sun (P″), and the Moon (P‴).

two expressions for the angular velocity associated with the Earth's spin, which must be equal,

$$\frac{2\pi}{D_{sid}} = \frac{2\pi + \gamma}{D_{sol}}.$$

Meanwhile, γ also represents the angle traversed by the Earth in its orbit during a solar day. Since the Earth orbits an angle of 2π in a sidereal year, we have the equality

$$\frac{\gamma}{D_{sol}} = \frac{2\pi}{Y_{sid}}.$$

Combining both expressions,

$$\frac{1}{D_{sid}} = \frac{1}{D_{sol}} + \frac{1}{Y_{sid}}. \tag{3.5}$$

Thus, the true sidereal day is somewhat shorter than the solar day (as is also borne out by the values mentioned earlier in this section). This is due to the fact that the Earth's spin has the same sense as its orbital motion, representing a prograde movement (if it were retrograde, then the sidereal day would be longer than the solar day).

We speak of the *true* sidereal day (D_{sid}) to distinguish it from another more commonly used "sidereal" day (D_{sid}^*) that refers to the time between two successive meridian transits of the *point of vernal equinox*, the point right above ♈ in Figure 3.8. It would be more apt to call D_{sid}^* a tropical day, but its traditional name is sidereal day, even though its point of reference is not a star. The difference from the true sidereal day is slight (see Table 3.2), being the result of the precession of the equinoxes, which causes the point of vernal equinox to drift slowly with respect to the stars. At the level of accuracy used for the tidal frequencies (see Tables 4.1 and 4.2), the distinction is however significant. As the sidereal day D_{sid}^* refers to the vernal equinox, its length can be calculated by using the tropical year instead of the sidereal year in (3.5):

$$\frac{1}{D_{sid}^*} = \frac{1}{D_{sol}} + \frac{1}{Y_{tro}}. \tag{3.6}$$

Finally, we note that all the lengths of time mentioned in this chapter – the various days, months, and years – are long-term *mean* values and as such they are very accurate. However, it is important to keep in mind that *individual* periods may vary significantly. For example, the length of the solar day varies by a range of 50 seconds, while individual lunar days may depart from their mean length by as much as a quarter of an hour. From here on, we will understand the "solar day" to be the *mean* solar day, without explicitly saying so, and likewise for the other periods.

3.2 Earth–Moon System

3.2.1 Orbit

In a similar way as the Earth moves around the Sun (Figure 3.1), the Moon moves around the Earth (Figure 3.5). In the Earth–Moon system, the center of mass lies at a distance of 4.7×10^3 km from the center of the Earth, i.e., within the Earth; this follows from the values listed in Table 3.1. The mean eccentricity ε of the Moon's orbit is 0.0549, but the actual eccentricity varies quite considerably due to perturbations by the Sun (see Section 3.2.4). In Figure 3.5, we only show the orbit of the Moon; it is understood that the Earth, too, traces a monthly orbit around the center of mass, though a much smaller one, as sketched in Figure 2.1.

The position closest to the Earth is called *perigee*; the opposite point is *apogee*. The time needed for the Moon to go from one perigee to the next is called the *anomalistic month*; it has a length of 27.55455 days (i.e., mean solar days). This

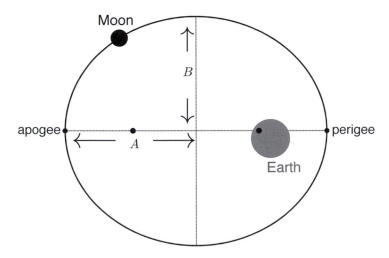

Figure 3.5 The Moon tracing an ellipse around the center of mass of the Earth–Moon system (black dot), which lies within the Earth. Meanwhile, the Earth traces a smaller orbit around the same point (not drawn). For visual clarity, the eccentricity of the ellipse is exaggerated.

period is slightly longer than the *sidereal month* of 27.32166 days, because the Moon's orbital ellipse turns slowly with respect to the stars (see Section 3.2.2).

Returning to Figure 3.4, we can relate the lunar day (D_{lun}) to the sidereal month (M_{sid}) and the true sidereal day. In a lunar day, the Moon progresses over an angle β with respect to its alignment to the stars, or expressed in terms of its orbital angular velocity,

$$\frac{\beta}{D_{lun}} = \frac{2\pi}{M_{sid}}.$$

In a true sidereal day, the Earth makes a full rotation over angle 2π, but in a lunar day, it makes an extra angle β (i.e., the passage from P′ to P‴), so

$$\frac{2\pi}{D_{sid}} = \frac{2\pi + \beta}{D_{lun}}.$$

Combining both expressions,

$$\frac{1}{D_{lun}} = \frac{1}{D_{sid}} - \frac{1}{M_{sid}}. \tag{3.7}$$

The period between alignments of the Sun, Earth, and Moon – the lunar cycle from full Moon to full Moon – is called the *synodic month*; its length is 29.53059 days. In a sidereal month, the Moon traces a full orbit with respect to the stars; it takes an extra orbital angle ν before the Sun, Earth, and Moon are again aligned

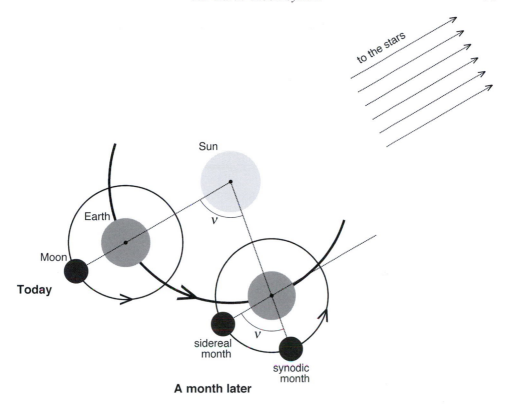

Figure 3.6 The Moon orbiting the Earth: after a sidereal month, the Moon has completed a full orbit and the Earth–Moon axis is again aligned to the same star. Around two days later, the axis regains its orientation with respect to the Sun, defining the synodic month.

(Figure 3.6). Expressed in terms of the Moon's orbital angular velocity, we have the equality

$$\frac{2\pi}{M_{sid}} = \frac{2\pi + \nu}{M_{syn}}.$$

In the course of a synodic month, the Earth has traversed an orbital angle ν, whereas it takes a sidereal year to make a full orbit, hence

$$\frac{\nu}{M_{syn}} = \frac{2\pi}{Y_{sid}}.$$

Combining both expressions, we find the equality

$$\frac{1}{M_{syn}} = \frac{1}{M_{sid}} - \frac{1}{Y_{sid}}. \tag{3.8}$$

Combining (3.5), (3.7), and (3.8), we obtain a simple relation connecting the lunar and solar days and the synodic month,

$$\frac{1}{D_{sol}} - \frac{1}{D_{lun}} = \frac{1}{M_{syn}}. \tag{3.9}$$

In tidal records, the synodic month manifests itself as the spring-neap cycle, reflecting the alignments of Earth, Sun, and Moon (Figure 2.8). The left-hand side of (3.9) indicates that we can alternatively interpret this cycle as a slow beat arising from the concurrence of the slightly different frequencies associated with the lunar and solar days.

Exercise

3.2.1 Show (with and without using formulas) that the number of solar days in a synodic month is precisely one higher than the number of lunar days.

3.2.2 Lunar Apsidal Precession

The Moon's orbital ellipse rotates slowly in its plane, a movement known as perigee shift or *lunar apsidal precession*; a full turn takes 8.847 years. Figure 3.7 shows part of the cycle.

Recalling the principle behind the spring-neap cycle, as outlined in Figure 2.8, we can readily see how the lunar apsidal precession affects that cycle. Let us first assume that the ellipse does not turn. During half of the year, full Moon would occur at the perigee side, and new Moon at the apogee side; as a result, spring tides would be stronger during full Moon than during new Moon. During the other half-year, it would be the other way round. This remains true if we take into account the apsidal precession (Figure 3.7), except that the dominance of full Moon (or new Moon) now lasts longer than half a year, namely 206 days. The upshot is that successive spring tides, which alternately occur near full Moon and new Moon, are unequal during most of the year, with the exception of situations B and D, when full and new Moon happen to have the same distance to the Earth.

We can derive this period as follows. Let Δt be the time interval from A to E in Figure 3.7, during which the ellipse has turned over angle α. In other words, Δt is the interval between successive alignments of the major axis with perigee directed to the Sun. A full turn of the ellipse with respect to the fixed stars takes 8.847 years, so we have the equality

$$\frac{\alpha}{\Delta t} = \frac{2\pi}{8.847 \text{ yr}}.$$

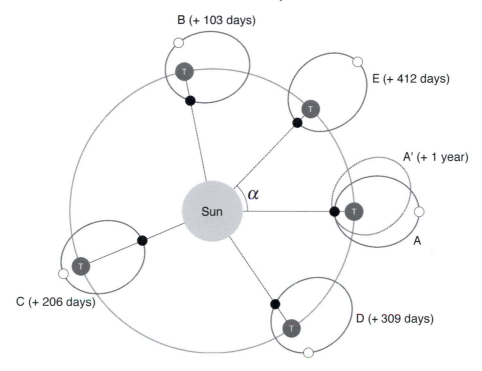

Figure 3.7 The changing orientation of the lunar orbital ellipse in the course of a year. Starting at A, we see how the ellipse slowly turns during the following stages (B,C,D). After one year, the Earth (T) is back to its original position, but the ellipse is now tilted (A′). It takes some extra time until the ellipse is again aligned to the Sun-Earth axis, as depicted in E. The phases of full and new Moon are shown as white and black circles, respectively. In A and E, new Moon coincides with perigee and full Moon with apogee; in C, it is the other way round.

In the same time Δt, the Earth has traversed one whole orbit plus angle α, while it takes a year to make an orbit, so

$$\frac{2\pi + \alpha}{\Delta t} = \frac{2\pi}{1 \text{ yr}}.$$

Hence the period Δt is 411.8 days. For the alignment of the major axis per se (without distinguishing perigees and apogees), the period takes half this value, i.e., 205.9 days, following the sequence A, C, E.

3.2.3 Inclination and Lunar Nodal Cycle

The Moon's orbit is inclined with respect to both the equatorial plane and the ecliptic, as sketched in Figure 3.8. The angle of the Moon's orbital plane with the *ecliptic* is nearly constant and always close to 5.1°. The interval between successive

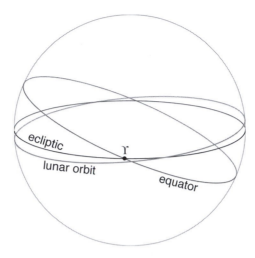

Figure 3.8 Sketch of the orbital plane of the Moon and its inclination with the equator and the ecliptic. Also indicated is the point of vernal equinox, ♈.

ascending crossings of the Moon through the ecliptic is the so-called draconic (or nodical) month of 27.21222 days. Similarly, we define the *tropical month* as the interval between two successive ascending crossings of the *equator* (the lunar equivalent of the solar vernal equinox), which has a length of 27.32158 days. The tropical month represents the lunar declinational period.[4]

Returning to Figure 3.4, we can replace the sidereal point of reference ("to the stars") by the point of vernal equinox. Correspondingly, replacing the angular velocities on the right-hand side of (3.7) by their tropical counterparts, we find another expression for the lunar day,

$$\frac{1}{D_{lun}} = \frac{1}{D_{sid}^*} - \frac{1}{M_{tro}}. \tag{3.10}$$

It is important to keep in mind the distinction between the *inclination* (κ), the angle of the lunar *orbital plane* with the equator, and the *declination* (δ), the angle of the Moon's *actual position* with the equator. The latter makes a monthly cycle and its maximum equals the inclination. The inclination, however, is not constant but varies itself over a cycle of 18.613 years, as the plane of the lunar orbit makes a precessional movement with respect to the *ecliptic* (Figure 3.9): the *lunar nodal cycle*. The *nodes*, the points where the Moon crosses the ecliptic, move slowly backward, which explains the difference in lengths between the nodical and tropical months. The angle with the ecliptic stays at all times close to 5.1°, but as a result

---

[4] The adjective tropical now refers to the turning of the *Moon*'s meridian transit at its highest or lowest point over the horizon.

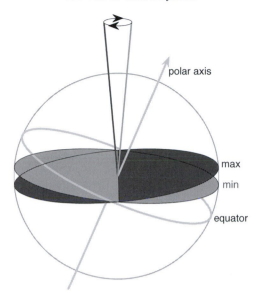

Figure 3.9 The 18.6-year lunar nodal cycle: the plane of the lunar orbit always makes an angle of 5.1° with the ecliptic (the horizontal plane in this graph, cf. Figure 3.8), but it slowly rotates backward in a precessional movement, as is also seen from the spin of the normal of the lunar plane. The states of minimum and maximum inclination are shown in light and dark gray, respectively.

of the precession, the angle with the equatorial plane varies greatly. In one phase of the nodal cycle the inclination is minimum

$$\kappa_{min} = 23.4° - 5.1° = 18.3°,$$

while 9.3 years later, in the opposite phase of the nodal cycle, it attains

$$\kappa_{max} = 23.4° + 5.1° = 28.5°.$$

These phases are illustrated in Figure 3.9 by the light and dark gray planes, depicting the lunar orbital plane at its extremes. A minimum inclination occurred in October 2015.

Exercise

3.2.1 This is a nocturnal exercise!

a) Estimate the length of the lunar day, following the idea of Figure 3.4.
b) Measure the angle of the Moon to the horizontal during its meridian transit (e.g., by measuring the length of the shadow of the Moon behind a stick or pencil, put upright on the floor). Call this angle σ.

c) Demonstrate that the declination is approximately given by

$$\delta = \sigma + \phi - 90°,$$

where ϕ is your latitude. You can check your result against values listed in online astronomical tables.

3.2.4 Other Lunar Variations

The Earth–Moon system is not autonomous; the Sun continually intervenes in their movements, especially in those of the Moon, which by virtue of its small mass is an easy prey for the Sun's attraction. The perturbations come in the form of periodic variations in the lunar orbital movement. In fact, two variations that we already discussed can be ascribed to the Sun's attraction: the lunar apsidal precession (8.847 years) and the lunar nodal cycle (18.613 years). They make the lunar orbital ellipse go around, but leave its shape unchanged. We now turn to variations involving the shape itself.

For the Earth–Moon system as a whole, the Sun's attraction is compensated by a centrifugal force, but this balance holds only at the center of mass of the Earth–Moon system. Like in the previous chapter, we are dealing with a tide-generating force, but now applied to the Earth–Moon system *as a whole*. In Figure 3.10, we have sketched the corresponding force.[5] Since the Earth almost coincides with

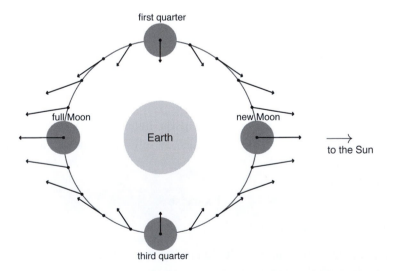

Figure 3.10 The Sun's tide-generating force acting on the Earth–Moon system.

---

[5] Here we take into account both the normal and tangential components of the tide-generating force (not just the tractive force).

the center of mass (cf. Figure 3.5), it is immediately clear that the effect will be relatively small for the Earth. Towards the peripheral Moon, on the other hand, the force gains strength. The direction in which the Moon is pulled depends on its phase: at new or full Moon, the solar tide-generating force pushes the Moon away from the Earth, whereas at first and third quarters, it pulls the Moon toward the Earth. In other words, if the Moon's orbit were a circle (as in Figure 3.10), the Sun's tide-generating force would turn it into an ellipse. Actually, the Moon's orbit already *is* an ellipse, and so it will depend on its orientation with respect to the Sun whether its eccentricity is enhanced or reduced. If the major axis is turned to the Sun (as in situations A, C, and E in Figure 3.7), the ellipse will be stretched and its eccentricity enhanced; if its minor axis is turned to the Sun (situations B and D, same figure), the minor axis will be stretched, thereby reducing the eccentricity. The period of this oscillation is 205.9 days; the eccentricity varies from about 0.026 to 0.078 in the course of this period.

Besides, there is another oscillation in the (instantaneous) eccentricity, known as *evection*, which has a monthly period of about 31.8 days (M_{eve}). This period is related to the lengths of the synodic and anomalistic months as

$$\frac{1}{M_{eve}} = \frac{2}{M_{syn}} - \frac{1}{M_{ano}}.$$

Finally, Figure 3.10 implies that the Sun augments the Moon's orbital speed toward full and new Moon, while reducing it toward the first and third quarters. This amounts to a modulation of the lunar orbital phase at a period of half the synodic month, known as *variation*.

3.3 List of Periods

We collect the main periods discussed in this chapter in Table 3.2. In the following chapter, we will use them to expand the tide-generating potential and extract the frequencies of the tidal constituents. Notice the analogies between Moon and Sun. The adjective *anomalistic* refers to the *elliptic* orbit, and the associated period comes as a month for the Moon and as a year for the Sun. The adjective *tropical* refers to their *declination*, again with monthly and yearly periods, respectively.

Further Reading

The subject matter of this chapter has a long history – traces of which can be seen in the sometimes peculiar nomenclature. Almost all the variations described in this chapter were already known, in one way or another, to ancient Greek astronomers (e.g., Hipparchos discovered the precession of the equinoxes in the second century BC). As the results have been long established, older texts retain much of

Table 3.2 *Periods related to the motions of the Earth and Moon. Units are mean solar hours, mean solar days, and mean tropical years.*

Symbol	Name	Time between two successive...	Mean length
D^*_{sid}	(common) sidereal day	meridian transits of the point of vernal equinox	23.934470 hours
D_{sid}	true sidereal day	meridian transits of a fixed star	23.934472 hours
D_{sol}	solar day	meridian transits of the Sun	24.000000 hours
D_{lun}	lunar day	meridian transits of the Moon	24.841202 hours
M_{ano}	anomalistic month	passages through perigee	27.55455 days
M_{tro}	tropical month	ascending lunar transits through equatorial plane	27.32158 days
M_{sid}	sidereal month	lunar orbital alignments to a fixed star	27.32166 days
M_{syn}	synodic month	phases of full Moon	29.53059 days
Y_{ano}	anomalistic year	passages through perihelion	365.2596 days
Y_{tro}	tropical year	ascending solar transits through equatorial plane	365.2422 days
Y_{sid}	sidereal year	terrestrial orbital alignments to a fixed star	365.2564 days
L_{aps}	lunar apsidal precession	sidereal alignments of the lunar major axis	8.847 years
L_{nod}	lunar nodal cycle	maximum lunar inclinations	18.613 years

their value. A good example is the lucid and informative treatise on astronomy by Russell et al. (1926) – from an age when global tidal dissipation was expressed in *horsepower*. More advanced mathematical treatments on the motions in the solar system, in particular lunar theory, can be found in Blanco and McCuskey (1961) and Fitzpatrick (2012). Details on basic aspects of celestial mechanics as outlined in Box 3.1 can also be found in advanced textbooks on classical mechanics, e.g., Goldstein (1980). We briefly discuss the calendar problem in Box 3.2; for a comprehensive historical overview on the subject, we refer to Richards (1998).

Box 3.2 **The Calendar Problem**

Of the celestial cycles, three stand out as the most apparent to the unaided eye:

- the solar day
- the synodic month (29.53059 days)
- the tropical year (365.2422 days)

The synodic month represents the period of lunar phases, from full Moon to full Moon; the tropical year represents one full cycle of the four seasons. Already in ancient cultures, life was organized around these three fundamental periods. They were known to great accuracy; the tropical year, for example, to the equivalent of two decimals in Babylonian times (c. 700 BC).

However, there was a problem: the three periods do not fit together. There is no whole number of solar days fitting in any whole number of synodic months, and likewise for synodic months versus tropical years. The periods are incommensurable. Counting solar days, the seasons would slowly drift through subsequent years, and the same is true for (religious) ceremonies associated with the lunar period. Ancient astronomers keenly noticed that an approximate match between the periods occurs over a 19-year cycle, the so-called *Metonic cycle* (after the Greek astronomer Meton, c. 430 BC, but possibly anticipated in Babylonian times). There is a near correspondence between 19 tropical years and 235 synodic months and 6,940 solar days. To accommodate the months, an extra month was intercalated in some years, so that one gets 12 years of 12 months and 7 years of 13 months, giving the total of 235 months in 19 years.

But in the long run, mismatches would still arise: the problem is fundamentally unsolvable. Indeed, in modern times, the problem has not been solved but evaded: our "month" has no true relation anymore with the lunar cycle; and in our "year," we occasionally have to add an extra day (in leap years) to keep it in sync with the seasons.

What does all this mean for the tides? The periods of the tidal constituents (to be discussed in the following chapter) are inherently related to these fundamental periods (as well as other celestial periods). Thus, the incommensurable periods feature in the tidal constituents, making the tidal signal as a whole an aperiodic phenomenon: the tide never exactly repeats itself.

4

Tidal Constituents and the Harmonic Method

4.1 Introduction

Equipped with the astronomical overview from the previous chapter, we return to the tide-generating potential (2.13). It contains time-dependent factors through distance r and declination δ, which vary periodically. These factors cause modulations of the fundamental diurnal and semidiurnal tides and produce long-period tides.

We adopt the convention that the periods from Table 3.2, denoted in capitals (D_{sol}, etc.), have their corresponding *frequencies* written by adding a tilde ($\tilde{D}_{sol}$, etc.). These frequencies will be expressed in degrees per hour, following the convention in the literature on tides; they are listed in Table 4.1.

Before entering into a more detailed analysis of the tide-generating potential, it is useful to have an intuitive understanding of the general principle by which tidal frequencies arise. Let us first consider the fictitious case of the Moon orbiting a circle in the equatorial plane. This gives us a semidiurnal tidal signal at frequency $2\tilde{D}_{lun}$ (i.e., a period of half the lunar day):

$$A_0 \cos 2\tilde{D}_{lun} t,$$

with a constant amplitude A_0. This is called the *principal lunar semidiurnal constituent*, M_2. We can think of it as a tide generated by a *fictitious celestial body* – following a notion introduced by Laplace: *astre fictif* – resembling the Moon in that it moves at the same mean orbital speed, but stripped of its declination and ellipticity.

We can come closer to reality if we allow the Moon's orbit to be an ellipse, which makes the amplitude of the tide vary at the period of the anomalistic month.[1] This would look like

---

[1] To keep things simple, we here vary only the amplitude, but in fact mechanical consistency requires that a time-dependent phase be added, since the orbital speed is higher at perigee and lower at apogee. This aspect is dealt with in Box 4.1.

Table 4.1 *Frequencies related to the motions of the Earth and Moon.*

Frequency	Associated period	Value (°/hour)
$\tilde{D}^*_{sid}$	(common) sidereal day	15.041069
$\tilde{D}_{sid}$	true sidereal day	15.041067
$\tilde{D}_{sol}$	solar day	15.000000
$\tilde{D}_{lun}$	lunar day	14.492052
$\tilde{M}_{ano}$	anomalistic month	0.544375
$\tilde{M}_{tro}$	tropical month	0.549017
$\tilde{M}_{sid}$	sidereal month	0.549015
$\tilde{M}_{syn}$	synodic month	0.507948
$\tilde{Y}_{ano}$	anomalistic year	0.041067
$\tilde{Y}_{tro}$	tropical year	0.041069
$\tilde{Y}_{sid}$	sidereal year	0.041067
$\tilde{L}_{aps}$	lunar apsidal cycle	0.004642
$\tilde{L}_{nod}$	lunar nodal cycle	0.002206

$$A_0(1 + 3\varepsilon \cos \tilde{M}_{ano}t)\cos 2\tilde{D}_{lun}t,$$

describing a *monthly modulation* of the basic M_2. The parameter ε represents the eccentricity of the ellipse (the reason for adding the factor of 3 will become apparent later, but is not essential at this point). Using formula (A.3), we expand this product as a sum of cosines,

$$A_0\cos 2\tilde{D}_{lun}t + \tfrac{3}{2}\varepsilon A_0 \cos(\overbrace{[2\tilde{D}_{lun} + \tilde{M}_{ano}]t}^{L_2}) + \tfrac{3}{2}\varepsilon A_0 \cos(\overbrace{[2\tilde{D}_{lun} - \tilde{M}_{ano}]t}^{N_2}).$$

This elucidates the general principle that monthly variations in amplitude, modulating the basic tide, give rise to *compound frequencies at the sum and difference frequencies*. As the modulation is slow compared to the basic frequency M_2, the compound frequencies $2\tilde{D}_{lun} \pm \tilde{M}_{ano}$ are still close to M_2 and thus semidiurnal. They are known as the semidiurnal lunar elliptic constituents N_2 and L_2.

Even though the compound frequencies are composed of frequencies with a real astronomical meaning (the lunar day and the anomalistic month), they are themselves devoid of any such meaning: no astronomical movement corresponds to the frequencies of N_2 or L_2. It is as if there were two more moons circling the equatorial plane, one going slightly faster than the Moon, the other, slightly slower. Even more than in the case of M_2, we can aptly regard them as *astres fictifs*. They are however a very useful fiction in that they allow us to treat the tide-generating potential as a simple sum of sinusoids instead of having to deal with a complicated expression involving many factors due to monthly (and yearly) variations. Each variation is effectively accommodated by adding yet another pair of *astres fictifs*, or as we will henceforth call them, tidal constituents.[2]

In the following sections, we derive the frequencies of the main constituents from the tide-generating potential.

4.2 Casting Celestial Motions into Sinusoids

The aim of this section is to derive approximations for the time dependence involved in the declinational cycles and elliptic orbits of the tide-generating bodies, the Moon and Sun.

4.2.1 Declination

We consider the declinational cycle of the tide-generating body for a given inclination κ (as defined in Section 3.2.3). For simplicity, we assume that the orbit is circular. In an orthogonal (x, y, z) frame of reference, we place the orbit in the horizontal xy-plane (Figure 4.1). The orbital motion can then be written as

$$X(t) = r \cos \Omega_{tro} t, \quad Y(t) = r \sin \Omega_{tro} t, \quad Z(t) = 0, \tag{4.1}$$

where r is the radius of the orbit. We are here dealing with the declinational cycles of the Moon and Sun, hence the period $2\pi / \Omega_{tro}$ stands for the tropical month M_{tro} in the case of the Moon, and for the tropical year Y_{tro} in the case of the Sun.

We denote the Earth's polar axis vector by $\vec{P}_{ax}$, which we place (without loss of generality) in the yz-plane. The tilt of the orbital plane of the tide-generating body to the equator is the inclination κ. The same angle turns up between the polar axis and the normal to the orbital plane, the latter being represented by the z-axis in our setting. Hence we can express the polar unit vector as

$$\vec{P}_{ax} = \begin{pmatrix} 0 \\ \sin \kappa \\ \cos \kappa \end{pmatrix}.$$

The declination δ is the latitude of the actual position of the tide-generating body (Figure 2.7), hence its angle with the polar axis is $\delta' = 90° - \delta$. Taking the inner product of the polar vector $\vec{P}_{ax}$ and the orbital position vector (X, Y, Z), we get

$$Y \sin \kappa = r \cos \delta'.$$

---

[2] They are also referred to as tidal harmonics or harmonic constituents. For a slightly different usage of the term "constituent," see Box 4.3.

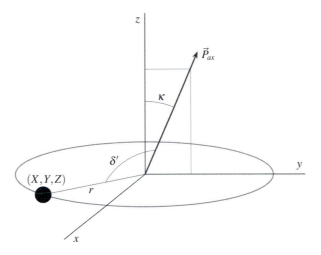

Figure 4.1 Sketch of the orbital plane of the tide-generating body, at (X, Y, Z), whose normal is inclined to the polar axis $\vec{P}_{ax}$ at angle κ.

With Y from (4.1), we thus find that the declination δ varies with time as

$$\boxed{\sin\delta = \sin\kappa \,\sin\Omega_{tro}t \,.} \tag{4.2}$$

4.2.2 Elliptic Orbit

We approximate the elliptic orbit by using the lowest-order solution of the Kepler problem, (3.2) and (3.4), on the assumption of small eccentricity ε. We thus find an expression for distance r as a a function of time:

$$r = A(1 - \varepsilon\cos\Omega_{ano}t), \tag{4.3}$$

where A is the semimajor axis. The period $2\pi/\Omega_{ano}$ refers to the time needed to traverse the ellipse: the anomalistic month M_{ano} in the case of the Moon, and the anomalistic year Y_{ano} in the case of the Sun.

Distance r enters the tide-generating potential (2.13) as an inverse cubic power. Using the binomial approximation (A.9), again for small eccentricity ε, we can approximate $1/r^3$ by

$$\boxed{\frac{1}{r^3} = \frac{1}{A^3}\,(1 + 3\varepsilon\cos\Omega_{ano}t + \cdots)\,.} \tag{4.4}$$

We have thus obtained a sinusoidal expression for the variation of $1/r^3$ (the so-called parallactic factor) with time.

4.3 Derivation of Tidal Constituents

In Chapter 2, we derived the tide-generating potential (2.13):

$$V = -\frac{3}{4}\left(\frac{a}{r}\right)^2 \frac{GM_m}{r} \times$$

$$\left[\overbrace{3(\sin^2\phi - \tfrac{1}{3})(\sin^2\delta - \tfrac{1}{3})}^{\text{long-period}} + \overbrace{\sin 2\phi \sin 2\delta \cos\Psi}^{\text{diurnal}} + \overbrace{\cos^2\phi \cos^2\delta \cos 2\Psi}^{\text{semidiurnal}} \right]. \quad (4.5)$$

We apply (4.5) to the Moon or Sun by choosing the parameters M (mass), r (distance), δ (declination), and Ψ (longitudinal phase) accordingly. For Ψ, the relevant period is the time interval between successive meridian transits of the Moon and Sun: the lunar day D_{lun} and solar day D_{sol}, respectively. The terms in (4.5) represent three species of tides: long-period, diurnal, and semidiurnal.

The long-period tides owe their existence entirely to the slow (bimonthly, biyearly, etc.) orbital variations in r and δ. The diurnal tides depend crucially on the presence of a declination (δ), which also acts as a modulating factor.

Latitude appears in different ways for the three species in (4.5). In the term for long-period tides, the factor $\sin^2\phi - 1/3$ reaches its maximum at the poles, vanishes at latitudes 35.3° N/S, and attains (minus) half its polar value at the equator. For diurnal tides, $\sin 2\phi$ vanishes at the equator and poles and has its maximum and minimum at 45° N/S. For semidiurnal tides, $\cos^2\phi$ nowhere changes sign and has its maximum at the equator and vanishes at the poles. Notice that this latitude dependence relates only to the *forcing* of tides and not to their actual appearance in the ocean.

For the derivation of the constituents, we need the astronomical periodicities listed in Table 3.2, restated as frequencies in Table 4.1. Notice that the difference between sidereal and tropical counterparts is always 0.000002° per hour; this difference originates from the precession of the equinoxes, whose cycle takes about 26,000 years. (Recall that the sidereal day D_{sid}^* is tropical in nature, in spite of its name: its point of reference is the vernal equinox.)

In the following sections, we derive the principal tidal constituents that emerge in the three species of tides: long-period, diurnal, and semidiurnal tides. We only consider lowest-order effects of declination and ellipticity, ignoring their combined effects. Moreover, we regard inclination κ as constant and defer a discussion of the role of the lunar nodal cycle to Section 4.3.4. We restrict ourselves to extracting just the *frequencies* of the main tidal constituents from (4.5); we do not derive the amplitudes of the constituents, for which it would be necessary to develop the tide-generating potential in a more systematic way (see the example in Box 4.1 and the references at the end of this chapter). For each constituent, we will state its name and symbol, in so far as they have one.

4.3.1 Long-Period Constituents

The time-varying factor in the long-period term in (4.5) is:

$$\left(\cdots \right) \frac{1}{r^3} (\sin^2 \delta - \tfrac{1}{3}), \tag{4.6}$$

where the dots stand for all the constant factors that are not needed here and that will henceforth be left out. Using the lowest-order approximations for the time dependence of the declination and the elliptic orbital movement, (4.2) and (4.4), we can write (4.6) as

$$\frac{1}{A^3} (1 + 3\varepsilon \cos \Omega_{ano} t) (\sin^2 \kappa \, \sin^2 \Omega_{tro} t - \tfrac{1}{3}).$$

With the identity (A.7), this becomes

$$\frac{1}{A^3} (1 + 3\varepsilon \cos \Omega_{ano} t) (\tfrac{1}{2} \sin^2 \kappa \, [1 - \cos 2\Omega_{tro} t] - \tfrac{1}{3}). \tag{4.7}$$

Solar

In the case of the Sun, the frequencies to be used in (4.7) are those corresponding to the anomalistic and tropical years, respectively.

Equation (4.7) thus contains a term with $\cos \tilde{Y}_{ano} t$, originating from the eccentricity of the Earth's orbit, which has a yearly period:

$$\tilde{Y}_{ano} \quad \longrightarrow \quad \mathbf{S}_a \text{ (solar annual elliptic)}.$$

In practice, the signal of this constituent is generally mixed with other nontidal yearly variations in sea level, related to seasonal effects in wind, atmospheric pressure and hydrology (i.e., the annual cycle in water storage on land).

Equation (4.7) also contains an oscillation at a half-yearly period, from the declinational term $\cos 2\tilde{Y}_{tro} t$. Hence

$$2\tilde{Y}_{tro} \quad \longrightarrow \quad \mathbf{S}_{sa} \text{ (solar semiannual declinational)}.$$

Lunar

In exactly the same way as for the Sun, but with years replaced by months, we find the two main lunar long-period constituents:

$$\tilde{M}_{ano} \quad \longrightarrow \quad \mathbf{M}_m \text{ (lunar monthly elliptic)}$$

and

$$2\tilde{M}_{tro} \quad \longrightarrow \quad \mathbf{M}_f \text{ (lunar fortnightly declinational)}.$$

The fortnightly constituent should not be confused with the spring-neap cycle, which also has a period of about 14 days, but depends on the Moon *and* Sun (Figure 2.8). In contrast, the Sun plays no role in the fortnightly constituent M_f. Moreover, the nature of the oscillation is entirely different. The M_f constituent truly is a fortnightly oscillation, whereas the spring-neap cycle is a fortnightly *modulation* of a semidiurnal oscillation. Finally, their periods are different: the fortnightly constituent's is 13.661 days (half a tropical month), whereas spring tides occur once every 14.765 days (half a synodic month).

4.3.2 Diurnal Constituents

The time-dependence of diurnal tides in the tide-generating potential (4.5) comes from the factor

$$\left(\cdots\right)\frac{1}{r^3}\sin 2\delta \cos \Psi.\tag{4.8}$$

Hereafter, we ignore the time-varying factor in $1/r^3$, being a higher-order effect that comes on top of the declinational modulation.

We need to cast the declinational factor $\sin 2\delta$ in terms of $\sin \delta$, since for the latter we have an expression in (4.2). Assuming a small inclination κ, and hence a small declination δ, we can make the following approximation:

$$\sin 2\delta = 2\sin \delta \cos \delta = 2\sin \delta \,(1 - \sin^2 \delta)^{1/2} = 2\sin \delta + \cdots.$$

For our purposes, this is an adequate approximation, as the leading term already illustrates the declinational effect in its simplest form. So, with the daily oscillation $\Psi = \tilde{D}t$, the time-varying part $\sin 2\delta \cos \Psi$ in (4.8) can be approximated by

$$2\sin \kappa \sin \Omega_{tro}t \cos \tilde{D}t.$$

Using (A.5), we thus obtain the compound frequencies

$$\tilde{D} \pm \Omega_{tro}.\tag{4.9}$$

Notice that the term (4.8) produces no "pure" diurnal frequencies $\tilde{D}$, only modulated ones.

Solar

For the Sun, the compound frequencies from (4.9) are

$$\tilde{D}_{sol} + \tilde{Y}_{tro} \quad \longrightarrow \quad \mathbf{K}_{1s} \text{ (solar declinational)}$$

and

$$\tilde{D}_{sol} - \tilde{Y}_{tro} \quad \longrightarrow \quad \mathbf{P}_1 \text{ (solar declinational)}.$$

The frequency K_{1s} corresponds exactly to that of the sidereal day $\tilde{D}^*_{sid}$, see Table 4.1. This is not coincidental. Choosing the tropical perspective, the solar day D_{sol} can be defined as the sidereal day D^*_{sid} plus the time needed to make up for the orbital movement around the Sun, whose speed is specified by the tropical year. Now, by adding the frequency of the tropical year to that of the solar day (which means: subtraction in terms of periods), we "undo" the orbital correction, as it were, which brings us back to the sidereal day. This is also evident from (3.6), which in terms of frequencies reads

$$\tilde{D}^*_{sid} = \tilde{D}_{sol} + \tilde{Y}_{tro}.$$

From this reasoning, it is a priori clear that *all* tide-generating bodies create a K_1 at the frequency of the sidereal day $\tilde{D}^*_{sid}$, provided that they have an inclination, of course.

Lunar

For the Moon, the compound frequencies from (4.9) are

$$\tilde{D}_{lun} + \tilde{M}_{tro} \quad \longrightarrow \quad K_{1m} \text{ (lunar declinational)}$$

and

$$\tilde{D}_{lun} - \tilde{M}_{tro} \quad \longrightarrow \quad O_1 \text{ (lunar declinational)}.$$

For the K_{1m} constituent, the same remark from the previous subsection applies; here we have from (3.10):

$$\tilde{D}^*_{sid} = \tilde{D}_{lun} + \tilde{M}_{tro}.$$

Since the frequencies of K_{1s} and K_{1m} are identical, they are usually taken together as

$$K_1 \text{ (luni-solar declinational)}.$$

It should however be noticed that the declinational correction stemming from the 18.6-year lunar nodal cycle (to be discussed in Section 4.3.4) applies only to the *lunar* part of K_1.

4.3.3 Semidiurnal Constituents

The time-varying factor in the semidiurnal term of (4.5) can be written as

$$(\cdots) \frac{1}{r^3} (1 - \sin^2 \delta) \cos 2\Psi. \tag{4.10}$$

Substituting the lowest-order approximation for the declination and the elliptic orbit, (4.2) and (4.4), we get

$$\frac{1}{A^3}(1 + 3\varepsilon \cos \Omega_{ano}t)(1 - \sin^2 \kappa \sin^2 \Omega_{tro}t) \cos 2\Psi.$$

Finally, we use (A.7) to rewrite the tropical factor,

$$\frac{1}{A^3}(1 + 3\varepsilon \cos \Omega_{ano}t)(1 - \tfrac{1}{2}\sin^2 \kappa [1 - \cos 2\Omega_{tro}t]) \cos 2\Psi. \tag{4.11}$$

This gives us three kinds of terms (if we ignore combinations of elliptic and declinational effects). First, the various constants in (4.11) produce a "pure" semidiurnal constituent $\cos 2\Psi$ whose frequency is not affected by elliptic or declinational effects:

$$2\Psi = 2\tilde{D}t. \tag{4.12}$$

Second, the ellipticity of the orbit creates a modulation of the semidiurnal tide, via the product

$$\cos \Omega_{ano}t \cos 2\tilde{D}t = \tfrac{1}{2}\cos([2\tilde{D} + \Omega_{ano}]t) + \tfrac{1}{2}\cos([2\tilde{D} - \Omega_{ano}]t).$$

We thus have the compound frequencies

$$2\tilde{D} \pm \Omega_{ano}. \tag{4.13}$$

Finally, declinational effects enter via the product

$$\cos 2\Omega_{tro}t \cos 2\tilde{D}t = \tfrac{1}{2}\cos([2\tilde{D} + 2\Omega_{tro}]t) + \tfrac{1}{2}\cos([2\tilde{D} - 2\Omega_{tro}]t).$$

This gives the compound frequencies

$$2\tilde{D} \pm 2\Omega_{tro}. \tag{4.14}$$

Solar

Applying (4.11) to the Sun, we use the frequencies of the anomalistic year, the tropical year, and the solar day. From (4.12) we thus have

$$2\tilde{D}_{sol} \quad \longrightarrow \quad \mathbf{S}_2 \text{ (principal solar semidiurnal)}.$$

This constituent has a period of exactly half a solar day, 12 h.

From (4.13), the elliptic constituents are

$$2\tilde{D}_{sol} + \tilde{Y}_{ano} \quad \longrightarrow \quad \mathbf{R}_2 \text{ (smaller solar elliptic semidiurnal)}$$

and

$$2\tilde{D}_{sol} - \tilde{Y}_{ano} \quad \longrightarrow \quad \mathbf{T}_2 \text{ (larger solar elliptic semidiurnal)}.$$

In practice, the constituent T_2 is more important than R_2. A more methodical analysis involving variations in phase speed explains this inequality in magnitude (see Box 4.1).

From (4.14), the declinational constituents are

$$2\tilde{D}_{sol} + 2\tilde{Y}_{tro} \longrightarrow \quad \mathbf{K}_{2s} \text{ (solar declinational semidiurnal)}$$

and

$$2\tilde{D}_{sol} - 2\tilde{Y}_{tro} \longrightarrow \quad \dots \text{ (solar declinational semidiurnal)}.$$

In this case, the former is the more important one; it has the period of half a sidereal day.

Lunar

For the Moon, the procedure is identical; we now use the frequencies of the anomalistic month, tropical month, and lunar day. Hence we find from (4.12):

$$2\tilde{D}_{lun} \longrightarrow \quad \mathbf{M}_2 \text{ (principal lunar semidiurnal)}.$$

This constituent has a period of exactly half a lunar day, 12 h 25 min 14 s.

From (4.13), the elliptic constituents are

$$2\tilde{D}_{lun} + \tilde{M}_{ano} \longrightarrow \quad \mathbf{L}_2 \text{ (smaller lunar elliptic semidiurnal)}$$

and

$$2\tilde{D}_{lun} - \tilde{M}_{ano} \longrightarrow \quad \mathbf{N}_2 \text{ (larger lunar elliptic semidiurnal)}.$$

Again, the elliptic constituents differ in amplitude, the latter being the more important one, as explained in Box 4.1.

From (4.14), the declinational constituents are

$$2\tilde{D}_{lun} + 2\tilde{M}_{tro} \longrightarrow \quad \mathbf{K}_{2m} \text{ (lunar declinational semidiurnal)}$$

and

$$2\tilde{D}_{lun} - 2\tilde{M}_{tro} \longrightarrow \quad \dots \text{ (lunar declinational semidiurnal)}.$$

Having the same frequencies, K_{2s} and K_{2m} are usually combined into the constituent K_2:

$$\mathbf{K}_2 \text{ (luni-solar declinational semidiurnal)}.$$

Exercises

4.3.1 In Section 4.3.2, effects of ellipticity were neglected. Derive the frequencies of the lunar diurnal constituents that arise from a lowest-order combination of elliptic and declinational effects.

4.3.2 Extract the frequencies of the main lunar tidal constituents from the last term in the higher-order tidal potential of Exercise 2.1.4 (i.e., the term $\cos\Theta$). In what sense are they different from the ones derived in this section?

Box 4.1 **Modulation of Speed**

In Section 4.1 we discussed how a modulation of the tidal amplitude, caused by a monthly variation in the Moon's distance, results in a pair of sinusoids at compound frequencies. However, as noted, this discussion is incomplete, because the variation in the Moon's distance also implies a variation in its orbital speed (Kepler's second law, see Box 3.1), which affects the phase of the tide. Effectively, this means that we can no longer treat the length of the lunar day as a constant $\tilde{D}_{lun}$, but have to take into account its monthly variation.

We first derive a convenient expression for the time-dependence of the orbital angle θ, which features in the second equation of (3.2). For small eccentricity ε, we can use the binomial series (A.9) to obtain an expression valid at order ε,

$$\cos\theta = \cos E - \varepsilon \sin^2 E + \cdots, \tag{4.15}$$

where the dots stand for higher-order terms in ε, which we neglect. Here E is the eccentric anomaly, which at order ε is given by E_2 in (3.4): $E = \omega t + \varepsilon \sin\omega t$. In the present case, the frequency ω of the orbital cycle is the anomalistic month, $\tilde{M}_{ano}$. Hence, by substitution of E in (4.15),

$$\cos\theta = \cos \tilde{M}_{ano}t - 2\varepsilon \sin^2 \tilde{M}_{ano}t + \cdots, \tag{4.16}$$

where we expanded the cosine and sine terms and retained only terms up to order ε, using (A.10) and (A.11). At this order of approximation, the right-hand side of (4.16) is identical to

$$\cos(\tilde{M}_{ano}t + 2\varepsilon \sin \tilde{M}_{ano}t),$$

so that

$$\theta = \tilde{M}_{ano}t + 2\varepsilon \sin \tilde{M}_{ano}t. \tag{4.17}$$

The time derivative of this expression is the angular speed $\dot{\theta} = \tilde{M}_{ano}(1 + 2\varepsilon \cos \tilde{M}_{ano}t)$, which attains its maximum at $t = 0$, which correctly corresponds to the Moon's perigee according to (4.3).

With the second term on the right-hand side of (4.17), we have obtained an expression for the monthly modulation of the lunar orbital phase. So instead of $\Psi = \tilde{D}_{lun}t$, the longitudinal phase becomes

$$\Psi = \tilde{D}_{lun}t - 2\varepsilon \sin \tilde{M}_{ano}t. \tag{4.18}$$

The minus sign accounts for the fact that the monthly frequency comes with a minus sign in the definition of $\tilde{D}_{lun}$, see (3.7). We now have the tools to round off the argument of Section 4.1. Starting from a basic tidal signal $A_0 \cos 2\tilde{D}_{lun}t$, we include the modulating effect of the orbital ellipticity on amplitude A_0 and on phase $2\tilde{D}_{lun}t$. Thus, with (4.4) and (4.18), the modulated signal becomes

$$A_0(1 + 3\varepsilon \cos \tilde{M}_{ano}t) \cos(2[\tilde{D}_{lun}t - 2\varepsilon \sin \tilde{M}_{ano}t])$$
$$= A_0(1 + 3\varepsilon \cos \tilde{M}_{ano}t) \left(\cos 2\tilde{D}_{lun}t + 4\varepsilon \sin 2\tilde{D}_{lun}t \sin \tilde{M}_{ano}t + \cdots \right)$$
$$= A_0 \cos 2\tilde{D}_{lun}t + 3\varepsilon A_0 \cos 2\tilde{D}_{lun}t \cos \tilde{M}_{ano}t + 4\varepsilon A_0 \sin 2\tilde{D}_{lun}t \sin \tilde{M}_{ano}t + \cdots$$
$$= A_0 \cos 2\tilde{D}_{lun}t + \tfrac{7}{2}\varepsilon A_0 \cos([2\tilde{D}_{lun} - \tilde{M}_{ano}]t) - \tfrac{1}{2}\varepsilon A_0 \cos([2\tilde{D}_{lun} + \tilde{M}_{ano}]t),$$

valid up to order ε. We have recovered the frequencies of the semidiurnal lunar elliptic constituents N_2 and L_2, but it is now clear that they have different amplitudes, N_2 being the larger of the two.

For declinational modulations, too, we can expect a difference in amplitude in the corresponding pair of constituents. In this case, the origin is not dynamical but comes from the projection of the orbit. For a Moon orbiting in the equatorial plane at constant speed, meridians are passed at a constant pace and hence longitudinal phase Ψ changes uniformly in time. However, if the orbit is tilted, then meridians are crossed more rapidly when the Moon stands high in the northern or southern hemisphere than when it crosses the equator, even though its orbital speed is constant. This means that the pace at which the longitudinal angle Ψ changes becomes variable. As in the previous example, a modulation of the phase renders the amplitudes of the pair of constituents unequal.

4.3.4 Correction for Lunar Nodal Cycle

In the previous section we have seen how modulations by monthly and yearly variations can be represented in terms of a number of constituents. We have however not yet taken into account the variations due to the 18.6-year lunar nodal cycle (which was introduced in Section 3.2.3). Because of its long period, it would introduce constituents whose frequencies depart only minutely from the basic ones. For this reason, the effects of the lunar nodal cycle have been traditionally included in a different way, namely by correcting *a posteriori* the amplitudes and phases of the lunar constituents.

For the amplitudes, this involves an adjustment factor. The M_2 is relatively weakly affected; its amplitude is modulated by $\pm 3.7\%$ during the lunar nodal cycle. However, the effect can be significant, since it is usually the largest constituent. The same percentage applies to N_2. In contrast, for K_2 the modulation is as large as $\pm 28.6\%$. (Here it has already been taken into account that the cycle affects only the lunar part of K_2.) Amplitudes of the lunar diurnal constituents vary by $\pm 18.7\%$ for O_1 and by $\pm 11.5\%$ for K_1. The fortnightly constituent M_f is relatively the most affected, by $\pm 41\%$. However, the resulting effect is small because the constituent itself is small.

In years when the lunar inclination is minimal (such as in October 2015), the lunar semidiurnal tides are stronger than usual, while the lunar diurnal tides are weaker. At the opposite phase of the lunar nodal cycle, it is the other way round. The modulation of the lunar constituents by the lunar nodal cycle has a significant effect on the tidal range and on the diurnal inequality. However, it leaves annual mean sea level unaffected since high tides are as much higher as low tides are lower, giving a cancellation in the mean (with the notable exception of intertidal areas, which lie above the low-tide level).

On the other hand, there is also a long-period nodal constituent, which has no effect on the tidal range but induces a small oscillation of annual mean sea level (a few centimeters at most), at a period of 18.6 years. In tide-gauge records, it is hardly possible to reliably extract this nodal signal, as it blends in with other (multi)decadal variabilities in sea level, mostly atmospherically induced.

4.4 Ranking of Constituents

The main constituents and their frequencies, derived in Section 4.3, are listed in Table 4.2, along with four constituents representing evectional and variational effects (see Section 3.2.4); the expressions for their frequencies indicate how they are composed.

In this chapter we focus on the *frequencies* of the constituents, but a fuller and more systematic development of the tide-generating potential allows their amplitudes to be calculated as well. In Figure 4.2, we have plotted the amplitude of the constituents from Table 4.2, along with 14 smaller ones, thus collecting the constituents that have an amplitude of at least 1% of that of the dominant M_2. They are neatly grouped into the three species of long-period tides, diurnal tides, and semidiurnal tides. Eight constituents stand out; they are, in order of importance: M_2, K_1, S_2, O_1, P_1, N_2, M_f, and K_2. Smaller ones are M_m, S_{sa}, and Q_1 (the last one finds its origin in a higher-order elliptic modulation of the lunar declinational O_1, see Exercise 4.3.1). The other constituents are barely visible.

This suggests that just a dozen or so constituents already represent the lion's share of the tidal potential. This is a correct inference, but a few qualifications

Table 4.2 *List of the main tidal constituents derived in Section 4.3, with five additional constituents (one diurnal and four representing evectional and variational effects). The Doodson numbers are explained in Box 4.3.*

Symbol	Species & name	Frequency	Value (°/hour)	Doodson numbers					
	Long-period constituents:								
S_a	solar annual elliptic	$\tilde{Y}_{ano}$	0.041067	0	0	1	0	0	-1
S_{sa}	solar semiannual declinational	$2\tilde{Y}_{tro}$	0.082137	0	0	2	0	0	0
M_m	lunar monthly elliptic	$\tilde{M}_{ano}$	0.544375	0	1	0	-1	0	0
M_f	lunar fortnightly declinational	$2\tilde{M}_{tro}$	1.098033	0	2	0	0	0	0
MS_m	evectional	$2\tilde{M}_{syn} - \tilde{M}_{ano}$	0.471521	0	1	-2	1	0	0
MS_f	variational	$2\tilde{M}_{syn}$	1.015896	0	2	-2	0	0	0
	Diurnal constituents:								
K_1	luni-solar declinational	$\tilde{D}^*_{sid}$	15.041069	1	1	0	0	0	0
P_1	solar declinational	$\tilde{D}_{sol} - \tilde{Y}_{tro}$	14.958931	1	1	-2	0	0	0
O_1	lunar declinational	$\tilde{D}_{lun} - \tilde{M}_{tro}$	13.943036	1	-1	0	0	0	0
Q_1	lunar elliptic declinational	$\tilde{D}_{lun} - \tilde{M}_{tro} - \tilde{M}_{ano}$	13.398661	1	-2	0	1	0	0
	Semidiurnal constituents:								
S_2	principal solar	$2\tilde{D}_{sol}$	30.000000	2	2	-2	0	0	0
M_2	principal lunar	$2\tilde{D}_{lun}$	28.984104	2	0	0	0	0	0
N_2	larger lunar elliptic	$2\tilde{D}_{lun} - \tilde{M}_{ano}$	28.439730	2	-1	0	1	0	0
L_2	smaller lunar elliptic	$2\tilde{D}_{lun} + \tilde{M}_{ano}$	29.528479	2	1	0	-1	0	0
K_2	luni-solar declinational	$2\tilde{D}^*_{sid}$	30.082137	2	2	0	0	0	0
ν_2	larger evectional	$2\tilde{D}_{lun} - 2\tilde{M}_{syn} + \tilde{M}_{ano}$	28.512583	2	-1	2	-1	0	0
μ_2	larger variational	$2\tilde{D}_{lun} - 2\tilde{M}_{syn}$	27.968208	2	-2	2	0	0	0

are in order. First, the tide-generating potential is one thing, but in the end we are more concerned with the actual tidal signal at a certain location. The complex way in which the tides propagate through the oceans and coastal seas (e.g., resonance in bays) may locally produce a different ranking of the constituents than in the tide-generating potential. Second, more important still is the local manifestation of tidal constituents that are not even present in the tide-generating potential,

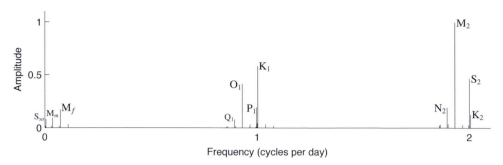

Figure 4.2 The amplitudes of the 28 most significant tidal constituents, derived from the tide-generating potential. They are normalized with respect to M_2. Based on values listed by Bartels (1957).

but arise from the hydrodynamical response to the astronomical constituents. In coastal seas and estuaries, nonlinearities due to advection and friction create their own myriad of tidal constituents, the *shallow-water constituents*, as multiples and compounds of the frequencies of the astronomical constituents; we discuss them further in Section 6.3. Some of the shallow-water compound frequencies are identical to astronomical ones, which further mixes up the ranking shown in Figure 4.2. Standard codes for tidal analysis typically contain more shallow-water constituents than astronomical ones. Third, the ocean tides are affected by their boundaries, the solid Earth below and the atmosphere above. Both experience tides in their own right, which leaves a fingerprint on ocean tides, in addition to effects of self-attraction and loading, as already briefly discussed in Section 1.5. The upshot is that the tide-generating potential, and its ranking of constituents, does not tell the whole story.

4.5 Harmonic Analysis

From a time-record of sea level ζ (or current velocity u) at a location of interest, one can estimate the amplitudes A_n and phases B_n of each tidal constituent, using a least-square fit. This means that the original signal is expressed as

$$\zeta = \sum_n \left[A_n \cos(\omega_n t - B_n) \right] + R, \tag{4.19}$$

where ω_n is the (known) frequency of constituent n. The remainder or residue is denoted by R. It partly contains nontidal effects, like storm surges. But R also represents other more fundamental limitations, notably any deviation from the idealized assumption that amplitudes and phases are constant. In reality, tides interact nonlinearly with storm surges, and the "constants" also change due to natural or manmade changes in the bathymetry in coastal environments.

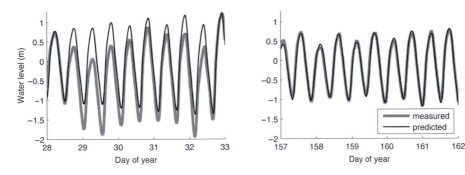

Figure 4.3 Two selected periods of observed water levels (gray line) from Figure 1.13, together with the predicted values (black line). Data supplied by the Dutch governmental agency Rijkswaterstaat.

By applying (4.19) to future times, we obtain the predicted signal. The exact procedure is however more complicated, for one also needs to take into account the time-dependent nodal factors in the amplitude and phases of the lunar constituents (see Section 4.3.4); this effect is included in standard codes. Tidal predictions are nowadays made routinely for many coastal locations worldwide, usually a year ahead. Examples are shown in Figure 4.3, where we have chosen two periods from the observed signal of Figure 1.13. In the panel on the left, we see that the observed level is mostly lower than the predicted one, sometimes by more than half a meter. This period corresponds to the one marked as box A in Figure 1.13, which is characterized by persistent southeasterly winds, causing a lowering of the sea level compared to the predicted tidal level. In contrast, in the panel on the right the correspondence between predicted and observed levels is very good, as this period had no strong winds.

To broadly characterize the local tide, the so-called form factor is often used, which is defined as a ratio involving the amplitudes (A) of four main constituents:

$$F = \frac{A_{K_1} + A_{O_1}}{A_{M_2} + A_{S_2}}. \tag{4.20}$$

We adopted the following common classification to make the global map in Figure 1.2: for $F < 0.25$, the tide is called semidiurnal; for $0.25 < F < 1.5$, mixed but mainly semidiurnal; for $1.5 < F < 3.0$, mixed but mainly diurnal; and for $F > 3.0$, diurnal.

4.6 Retrieving the Celestial Harmonies

In Figure 1.13 we showed a year-long record of sea level from a tide gauge at the North Sea coast. From this record we can extract the tidal constituents that we derived in Section 4.3. We restrict ourselves to the main constituents, which we first

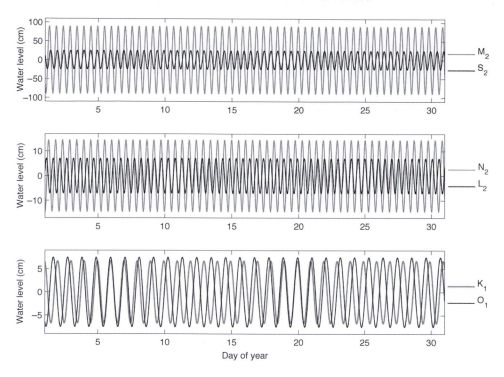

Figure 4.4 The main constituents from the tide gauge record shown in Figure 1.13, plotted for January 2014. Notice the difference in vertical scale between the panels.

show separately in Figure 4.4. By far the largest is the principal lunar semidiurnal constituent M_2; the second largest is the principal solar semidiurnal constituent S_2 (upper panel). In the middle panel, we show the lunar constituents associated with the eccentricity of the lunar orbit, N_2 and L_2; they represent the monthly variation in the distance of the Moon. Finally, in the lowest panel, we show the diurnal constituents associated with the declination of the Moon, K_1 and O_1.

In fact, K_1 contains solar effects as well, as explained in Section 4.3.2. Furthermore, the constituent L_2 seems relatively large compared to N_2, set against the result from Box 4.1. This discrepancy is caused by the presence of shallow-water constituents, one of which replicates the frequency of L_2 and enhances its signal at the expense of N_2, as further discussed in Section 6.3. However, so far as the purpose of this section is concerned – visualizing the principal astronomical periodicities in the tide, these aspects can for now be ignored.

We first leave out the Sun and focus on the elliptic effects of the Moon's orbit, ignoring also the declination for the moment. Thus, what we see in Figure 4.5 broadly represents the tidal signal the Moon would produce if it lay in the equatorial plane and if no Sun were present. The tidal range varies substantially in the course of the anomalistic month, due to the Moon's elliptic orbit.

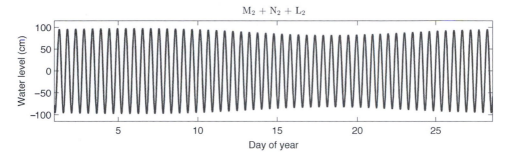

Figure 4.5 Superposition of constituent M_2 with the main constituents representing lunar elliptic effects, N_2 and L_2. The time span is one anomalistic month, the period associated with the Moon making a full elliptic orbit.

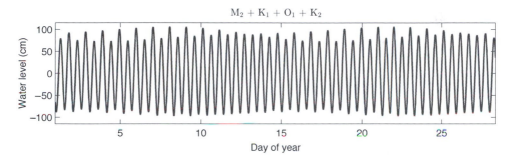

Figure 4.6 Superposition of constituent M_2 with the main constituents representing declinational effects, K_1, O_1, and K_2. The time span is one tropical month, the period pertaining to the lunar declinational cycle.

Ignoring now elliptic effects, we focus instead on the declination and superpose M_2 with the diurnal constituents K_1 and O_1 (and K_2, which is also due to declination). The result is shown in Figure 4.6. There is now a *diurnal inequality*: high tides are alternately higher and lower. But the diurnal inequality is not permanently present: it disappears twice during the tropical month (around days 13 and 27). During this period, the Moon crosses the equatorial plane twice, which implies a vanishing declination, and this bimonthly character is reflected in the tidal signal in Figure 4.6.

In the previous two figures we isolated the Moon as we zoomed in on the elliptic and declinational effects in the lunar tidal signal. We now bring the Sun back into the equation and first show the obvious but important result of the spring-neap cycle, obtained by superposing M_2 and S_2 (Figure 4.7). The spring-neap cycle occurs at a period of half a synodic month (cf. Figure 2.8) and reflects the phases of the Moon, as indicated by the circles in Figure 4.7. At this location spring tides occur about two days after new or full Moon.

The spring-neap cycle in Figure 4.7 is entirely regular, but we may expect that the variation of the Moon's distance (elliptic effects) will perturb this picture. Thus

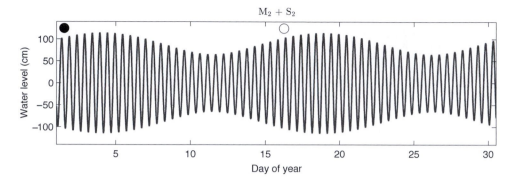

Figure 4.7 Spring-neap cycle: the superposition of constituents M_2 and S_2, covering a synodic month. The moments of conjunction with the Moon are also indicated: new Moon (black circle) and full Moon (open circle).

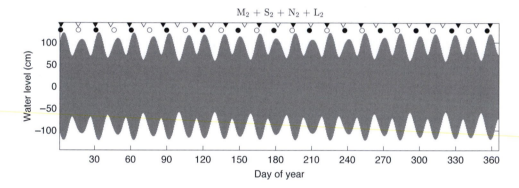

Figure 4.8 The perturbed spring-neap cycle, due to elliptic variations by N_2 and L_2. Phases of new Moon (black circle) and full Moon (open circle) are indicated, as well as the stages of lunar perigee (black triangle) and apogee (open triangle).

adding N_2 and L_2 to Figure 4.7 (or, equivalently, adding S_2 to Figure 4.5), we illustrate the resulting signal for a full year (Figure 4.8). First of all, the spring-neap cycle has become uneven; successive spring tides now generally differ in strength. At the beginning of the year, highest spring tides occur at New Moon; later in the year, this changes to Full Moon. With Figure 3.7 in mind, we can readily understand what causes this cycle: it reflects the cycle of perigee and apogee (indicated by triangles in Figure 4.8) in combination with a slow turning of the lunar orbital ellipse (the 8.85-year cycle of lunar apsidal precession, Section 3.2.2). Early in the year, new Moon coincides more or less with perigee and full Moon with apogee (situation A in Figure 3.7). Hence tidal ranges are larger at new Moon than at full Moon. However, as the anomalistic month is slightly shorter than the synodic month (i.e., the succession of triangles is slightly more rapid than the succession of circles

Figure 4.9 Sketch of the orientation of the lunar orbital plane, with respect to the position of the Sun and the polar axis, during summer and winter (in the northern hemisphere, or NH).

in Figure 4.8), the situation changes gradually, and at around day 110, successive spring tides become nearly equal (situation B in Figure 3.7). Subsequently, phases of full Moon and perigee come increasingly close, resulting in higher spring tides during full Moon than during new Moon, most clearly so around day 220 (situation C in Figure 3.7). Figure 4.8 thus offers a vivid illustration of the interplay between the different monthly cycles.

Finally, we return to a feature in the diurnal inequality that we identified in Figure 1.14, namely that higher high tides occur at daytime during summer and at nighttime during winter. This phenomenon can be understood by looking at the orientation of the three planes involved, as illustrated in Figure 4.9. The lunar orbital plane and the ecliptic are tilted in the *same* sense with respect to the equatorial plane. The implication is that, during summer, the lunar orbital plane intersects the northern hemisphere at daytime, producing a higher high tide. During winter, the intersection occurs at nighttime. In the southern hemisphere, both the seasons and the intersections are reversed, and as a result, the rule remains the same.

4.7 Timing of the Tide

The very word *tide* is etymologically related to the notion of time (as is more obvious in the Germanic languages). To explore this relation in an empirical way, we return once again to the tide-gauge record from a station in the North Sea, shown in Figure 1.13. In Section 4.6, we found that the semidiurnal lunar constituent M_2 is the largest constituent and thus dominates the signal at this location. We now examine the timing of low and high tides, and the associated tidal period.

4.7.1 Cycles in the Phasing of the Tide

By way of example, we have extracted the moments of low tides from the tide-gauge record of Figure 1.13; they are shown in Figure 4.10. The moments shift day by day, producing a periodic pattern of diagonals that is neatly phase-locked to the

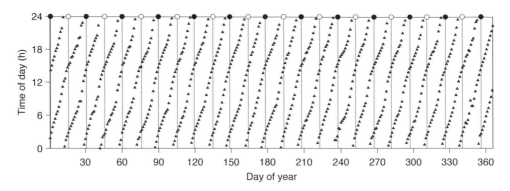

Figure 4.10 The moments of low tides during the day (in hours), plotted for the year 2014. Derived from the tide gauge record of Figure 1.13. Also indicated are the lunar phases of full Moon (open circle) and new Moon (closed circle), with their timing in vertical lines.

lunar cycle. At new and full Moon, the first low tide of the day at this location always occurs at around 3 o'clock in the morning.

There thus appears to be a semi-synodic cycle in the timing of low tides, but this is not caused by the presence of solar constituents like S_2. In fact, the cycle would occur even if the tidal signal were defined by M_2 alone. Here, the synodic month comes into play not via solar constituents but via our convention of using the *solar* day as the unit of time. The M_2 period is incongruent with this unit, hence the daily shift in the occurrence of low tides. A low tide at a certain moment of the day is shifted by an average of about 50.5 minutes the next day, reflecting the difference between the lunar and solar days (as illustrated in Figure 3.4). After a sequence of 28.5 instances, the shifts add up to a full solar day, so we are back to where we started in terms of the timing with respect to the solar day. Meanwhile, a period of 28.5 lunar days, or 29.5 solar days, has passed – a synodic month. The same happens for the other concurrent sequence of low tides (the adjacent diagonal), which effectively reduces the interval of the periodicity to half a synodic month in the timing of low tides.

4.7.2 The Tidal Period – What Is It?

In coastal areas, large amounts of water and suspended sediment (and other constituents) are flushed back and forth with ebb and flood. What often interests us most is the *net* effect of this cycle. This means that we have to integrate the instantaneous transport over a tidal period. But exactly what is its length?

It is apparent from Figure 4.10 that the diagonals are not exactly straight lines. They show wiggles and little bends, meaning that the vertical interval from one

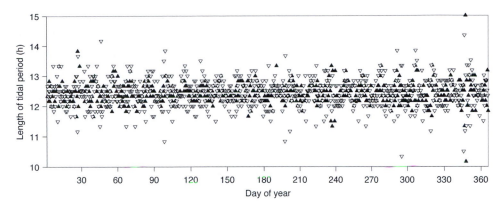

Figure 4.11 The tidal period calculated as the interval between two successive high tides (open triangles) and between two successive low tides (closed triangles). Derived from the tide gauge record of Figure 1.13.

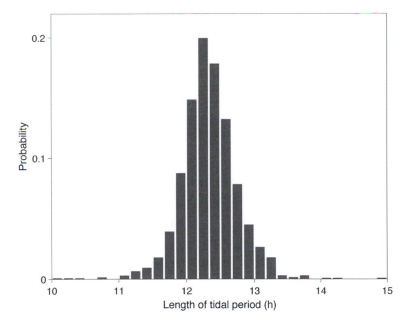

Figure 4.12 Histogram of all tidal periods from Figure 4.11, with normalized probability (the total sums to one).

point to the next (on the same diagonal) is not uniform. In particular, the interval is generally shortest a few days after full or new Moon. This already indicates that the tidal period has a certain unsteadiness. This is further illustrated in Figure 4.11, where we have plotted the time intervals between successive low tides and between successive high tides. Together, they are collected in a histogram in Figure 4.12.

The tidal period shows a considerable spread, implying that there is actually no such thing as "the" tidal period. At any one day, different definitions (e.g., using the interval between successive low tides, or between successive high tides, or between alternate slack waters) may produce different results. Sticking to one definition, values vary through time. This variability can be ascribed to the presence of a great number of tidal constituents, each with its own well-defined frequency, but their superposition produces an irregular period for the signal as a whole. This problem is compounded by the occasional presence of storm surges or depressions. For example, the interval between successive high tides can be prolonged if the second one is delayed by a mounting storm surge.

Amid all this variability, there is a remarkable constant hidden in Figure 4.11. If we calculate the annual mean tidal period, using the periods based on successive high or low tides, both produce the value of 12 h 25 min 14 s, which is – to the second precise! – the period of the semidiurnal lunar constituent M_2. This is not coincidental. It is a mathematical property of a sum of sines of different periods, that the long-term mean period converges to the period of the largest constituent, which at this tide gauge station is M_2 (see Box 4.2).

Box 4.2 Long-Term Mean Tidal Period

Here we examine the long-term mean tidal period for a superposition of two constituents (numerically, it could of course be done for many more); we demonstrate that it equals the period of the larger constituent. We start with a water level ζ described by

$$\zeta = \cos(\omega_a t) - \varepsilon \cos(\omega_b t), \tag{4.21}$$

with frequencies $\omega_{a,b}$. We assume that the second constituent is relatively small, $\varepsilon \ll 1$.

First of all, we need to find the moments of high and low waters. Setting the derivative $d\zeta/dt$ equal to zero, we obtain

$$\sin(\omega_a t) = \varepsilon \mu \sin(\omega_b t), \tag{4.22}$$

where $\mu = \omega_b/\omega_a$. We take the arcsine of both sides of (4.22), giving

$$\omega_a t = \arcsin[\varepsilon \mu \sin(\omega_b t)]. \tag{4.23}$$

Since the argument of the arcsine is much smaller than one, we can use the Taylor series $\arcsin x = x + \cdots$ and neglect higher order terms. Taking into account the variety of angles that would produce the same sine (i.e., the part in square brackets), we can split the solution in two cases:

$$\omega_a t = \begin{cases} \varepsilon \mu \sin(\omega_b t) + 2\pi n + \cdots \\ \pi - \varepsilon \mu \sin(\omega_b t) + 2\pi n + \cdots, \end{cases} \tag{4.24}$$

with $n = 0, 1, 2, 3 \cdots$. It can be verified numerically that the step from (4.23) to (4.24) involves only a small error even for fairly large values $\varepsilon < 1$, so the approximation is not very restrictive. Equations (4.24) have the same formal structure as the Kepler problem (3.1) and can be solved by the methods mentioned in Box 3.1. However, we do not actually need the solution of (4.24) to arrive at our result.

By substitution of (4.24) in (4.21), we find that the first equation in (4.24) corresponds to the moments of high water, and the second equation to the moments of low water. Both cases can be dealt with similarly; here we consider the case of high waters in further detail.

The first equation in (4.24) implicitly defines the moments t_n of high water. The time between two successive high waters, the tidal period $T_n = t_{n+1} - t_n$, is given by

$$\omega_a T_n = \varepsilon \mu [\sin(\omega_b t_{n+1}) - \sin(\omega_b t_n)] + 2\pi . \tag{4.25}$$

Summing a stretch of N successive periods, we have

$$\omega_a \sum_{n=1}^{N} T_n = \varepsilon \mu [\sin(\omega_b t_{N+1}) - \sin(\omega_b t_1)] + 2\pi N . \tag{4.26}$$

Hence the mean tidal period $\langle T \rangle$ is given by

$$\langle T \rangle = \frac{1}{N} \sum_{n=1}^{N} T_n = \varepsilon \mu \frac{\sin(\omega_b t_{N+1}) - \sin(\omega_b t_1)}{\omega_a N} + \frac{2\pi}{\omega_a} , \tag{4.27}$$

which for large N tends to $2\pi / \omega_a$, the period of the larger constituent.

4.7.3 Circa-Tidal Clocks in Marine Life

In coastal regions, intertidal areas provide a special ecological habitat. They are flooded during high tides and fall dry during low tides, so marine organisms living there need to be able to cope with the twice-daily radical change in their environment. For example, crabs tend to hide in burrows during low tides, only to leave them during high tides to roam the seafloor. Diatoms migrate downward into the sediment layer during high tides to be protected from strong currents, and back during the daytime low tide for photosynthesis. These are just a few examples out of many, including annelids (worms), mollusks, and crustaceans, where physiological and behavioral rhythms of marine life are influenced by the tidal cycle. It seems natural enough that these species would respond to external cues, like currents or hydrostatic pressure, from which they sense the phase of the tidal cycle.

However, marine biologists have investigated what happens when they are removed from their natural habitat and placed in a laboratory where no external cues are present (constant light, no tides). Surprisingly, it turns out that many species continue to follow their daily and tidal rhythms, for weeks or even longer

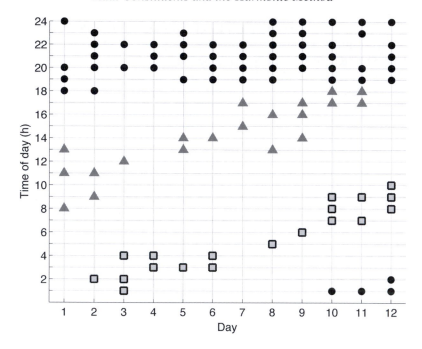

Figure 4.13 The timing of locomotor activity of a crab, after it was brought from its habitat into a laboratory setting without natural external cues. Two tidal rhythms can be distinguished (squares and triangles), which on average exhibit daily shifts by approximately a lunar day, while a steady solar rhythm (circles) is followed as well. Data from Palmer (1996, figure 4).

(Figure 4.13). This means that they somehow have an internal biological clock, or perhaps even several clocks, to follow the light/dark rhythm of the solar day as well as the daily shifts in each of the low tides, reflecting the lunar day. The advantage of a clock is that the organism can physiologically anticipate the upcoming events. In the case of tidal clocks, organisms encounter the same problem that we discussed in the previous section, namely that the tidal period is quite variable. Hence biologists speak of circa-tidal clocks. There is no evolutionary advantage in having a very precise clock. In fact, some individuals of a population may dispense with it altogether. An intriguing question is where the clock resides and how it functions on a molecular level. This is still an area of active research.

Further Reading

In Section 4.3, we essentially followed the approach of Doodson and Warburg (1941, chapter VI), with an emphasis on an intuitive understanding of the origin of the tidal constituents, rather than on a systematic development of the tide-generating potential. A description of the latter, which yields not only the

---

Box 4.3 **Doodson Numbers**

In a systematic development of the tide-generating potential, as proposed by Doodson (1921), the time dependence of distance r, declination δ and longitude Ψ is expanded into sinusoids using six fundamental arguments:

$$\tau = \tilde{D}_{sol}\, t + h - s$$
$$s = s_0 + \tilde{M}_{tro}\, t + \cdots$$
$$h = h_0 + \tilde{Y}_{tro}\, t + \cdots$$
$$p = p_0 + \tilde{L}_{aps}\, t + \cdots$$
$$N = N_0 - \tilde{L}_{nod}\, t + \cdots$$
$$p' = p'_0 + (\tilde{Y}_{tro} - \tilde{Y}_{ano})\, t + \cdots .$$

Here t is expressed in solar hours. In the literature, time is usually expressed in Julian centuries in s, etc., but to simplify the comparison with our nomenclature, we use solar hours instead. At the chosen starting date ($t = 0$), mean longitudes of the Moon (s), Sun (h), lunar perigee (p), the ascending lunar node (N), and perihelion (p'), have initial phases s_0, etc. Their values are known from astronomical models. Here longitude is measured along the ecliptic from the vernal equinox eastward. The dots stand for relatively small quadratic terms in t, which we leave out here. The longitude of perihelion p' represents the secular shift through the tropical year of the moment of perihelion, which results from a combination of the precession of the equinoxes and the perihelion shift, and has a period of about 21,000 years.

By taking time derivatives, we readily recover the familiar expressions from Chapter 3. In particular, in $\dot{\tau} = \tilde{D}_{sol} + \tilde{Y}_{tro} - \tilde{M}_{tro}$ we recognize the expression for the lunar day $\tilde{D}_{lun}$, combining (3.6) and (3.10). Arguments of all the other tidal constituents can likewise be expressed as a sum of multiples of the fundamental ones (τ, s, h, p, N, p'). For example, the larger lunar elliptic constituent N_2 can be written as $2\tau - s + p$, whose time derivative corresponds to $2\tilde{D}_{lun} - \tilde{M}_{ano}$ (see Table 4.1), and thus has multiplication factors $(2, -1, 0, 1, 0, 0)$: the *Doodson numbers*. For the main constituents, they are listed in Table 4.2. Frequencies that have the same first three Doodson numbers are very close and are often together designated as a "constituent."

---

frequencies but also the amplitudes and phases of the constituents, can be found elsewhere: first of all in Doodson (1921), later revisited by Schureman (1940) and Bartels (1957), and updated by Cartwright and Taylor (1971) and Cartwright and Edden (1973). These expansions already provide more astronomical constituents than are practically needed for tidal predictions. Box 4.3 provides a brief description of the standard notation adopted in the literature.

A tidal prediction of momentous importance was the one prior to the Normandy amphibious landing in World War II (D-Day). As Parker (2011) explains in an engrossing recount, the success or failure of the operation – in a region with tidal ranges as large as six meters – depended crucially on getting the timing of high and low tides right.

In Section 4.6, we used the code T_TIDE, developed and described by Pawlow-icz et al. (2002), with more background information in Godin (1972). More details on the nodal correction factors, and on tidal prediction in general, can be found in Pugh and Woodworth (2014). Ideas to improve tidal predictions do not rely on including more constituents (which would not help, since their number is already adequate) but on tweaking the method. For example, Harris (1991) demonstrates that the tidal constants (amplitudes and phases) have a certain sensitivity to the starting date of the record, and suggests that this problem is reduced if one takes a record length of 354.5 days (approximately 12 synodic months) instead of the common (tropical) year. Hibbert et al. (2015) propose a method to include empirical correction factors for high waters during storm surges. Egbert and Ray (2017) provide a comprehensive overview on global and regional tidal prediction based on models, satellite altimetry, and data assimilation.

The long-period tides form a class apart: although of little importance from the perspective of tidal prediction, they are of fundamental interest: for example, regarding the long-standing question to what extent they follow the *equilibrium tide*, i.e., form a quasi-static response to the first term of the tide-generating potential (4.5). For the fortnightly and monthly constituents M_f and M_m, significant departures from the equilibrium tide have been found (in the form of radiation of Rossby waves). A good place to start is the paper by Kantha et al. (1998), see also Egbert and Ray (2003a) on M_f. For the 18.6-year lunar nodal cycle, the question has still not been fully resolved; Woodworth (2012) presents a self-consistent form of the equilibrium tide by including the effects of self-attraction and loading. A special kind of long-period tide is the *pole tide*, which arises from a nutational movement of the polar axis (the Chandler wobble), at a period of about 14 months. Desai (2002) gives a detailed description of the pole tide, both on observations and theory.

The various astronomical cycles imply a periodic strengthening and weakening of the tides, as illustrated in Section 4.6. Since the periods of these cycles are different, the maxima occasionally add up to form *peak tides*, for example when spring tides coincide with perigee and perihelion. For a period of 3,000 years, Ray and Cartwright (2007) calculated and listed these events and unraveled the causes of the principal variations and long-term trends. A list of contemporary peak tides is provided by Pugh and Woodworth (2014, table 3.4). From a practical point of view, the importance of peak tides should not be exaggerated, for their effect is usually masked by local atmospheric conditions.

The problem of how to define the tidal period, and its consequences for calculating tide-averaged currents, is further explored in Duran-Matute and Gerkema (2015). Finally, more on biological clocks in marine organisms can be found in the reviews by Palmer (1996) and Tessmar-Raible et al. (2011).

5

Tidal Wave Propagation

In previous chapters we focused on the tide-generating potential and on the various periods (tidal constituents) that emerge from it. The tide-generating potential works across the oceans. The resulting tidal signal at any location is a blend of signals generated elsewhere, coming from areas in the world oceans where the tide-generating force injects energy into the system (for semidiurnal tides, the dominant areas of input are the South Atlantic Ocean and the western part of the Indian Ocean). The step from the global tide-generating potential to the local response is complex and can only be dealt with in numerical tidal models. Nonetheless, despite the complex connection between forcing and response, the resulting tidal waves are globally organized in orderly patterns. Key features in them can already be understood by examining the properties of tidal wave propagation in basic hydrodynamic models, which forms the subject of this chapter.

5.1 Introduction

On the basis of observations from satellite altimetry, detailed and accurate global maps can nowadays be made of the amplitude and phase of the surface elevation of individual constituents. Conceived as a sinusoid as in (4.19), each constituent n can be expressed as

$$A_n(\lambda, \phi) \cos(\omega_n t - B_n(\lambda, \phi)).$$

For selected frequency ω_n, spatial images of amplitude A_n and phase B_n can be produced.[1] Here λ is longitude and ϕ latitude.

In Figures 5.1 and 5.2, we show maps of the most important semidiurnal and diurnal constituents, M_2 and K_1, respectively. For both, the presence of the continents plainly shapes the pattern of amplitudes and phase propagation. In particular,

---

[1] Obviously, such maps can only be made for individual constituents, but not for the tidal signal as a whole.

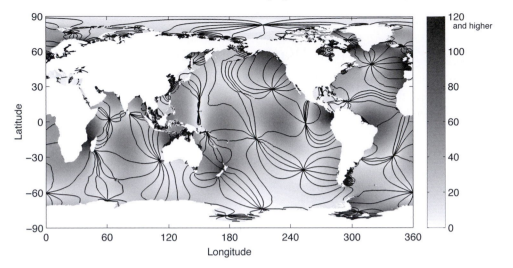

Figure 5.1 Global map of the M_2 constituent, with amplitude in centimeters. Co-phase lines (i.e., lines connecting points of equal phase) are shown in black. Lags between neighboring lines represent a phase difference of 1/12 of the M_2 period (i.e., slightly more than one solar hour). Figure generated using Aviso+ products, courtesy of LEGOS/Noveltis/CNES/CLS.

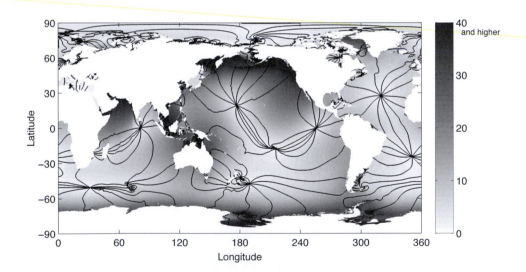

Figure 5.2 Same as Figure 5.1, but now for K_1. Lags between neighboring lines here represent a phase difference of 1/12 of the K_1 period (i.e., slightly less than two solar hours). Figure generated using Aviso+ products, courtesy of LEGOS/Noveltis/CNES/CLS.

the highest amplitudes always occur near the continents. Furthermore, co-phase lines border the continents perpendicularly, or nearly so. High and low tides thus propagate fundamentally *along* the continents, not toward them.

If we follow the co-phase lines away from the coast, we observe that they come together at certain spots, which are called *amphidromic points*. At these points, the amplitude vanishes. This is a mathematical necessity, since if there were a signal, it would have to exhibit at once all the different phases that come together at this point, which is impossible. The sense of propagation varies, but counterclockwise rotation around amphidromic points is generally found in the northern hemisphere, and clockwise rotation in the southern hemisphere.

We have so far focused on the common features of Figures 5.1 and 5.2, but their global patterns are clearly very dissimilar: high amplitudes occur in different places and so do amphidromic points. Rather than a one-pattern-fits-all approach, semidiurnal and diurnal constituents need to be treated separately, each with their own basin-wide characteristics. Meanwhile, semidiurnal and diurnal constituents coexist in the tidal signal as a whole, so if one of them has an amphidromic point at a certain location, there is still a tidal signal, because the other has not.

Simple models for tidal wave propagation cannot capture the full complexity of these global maps, but it turns out that they can nonetheless explain some of their characteristics, notably the occurrence of amphidromic points and the role of continental slopes as a wave guide.

Continental slopes form the boundaries of ocean basins, as illustrated in Figure 5.3, where we describe the principal features of the ocean floor. At each side of the basin, a continental slope marks the transition between the deep ocean and a shallow continental shelf sea. In addition, the deep basin features a large mid-ocean ridge, which runs as one long chain through all ocean basins (i.e., Atlantic, Indian, and Pacific Oceans). This is the area where new ocean floor is formed, as magma wells up from the mantle, and where the tectonic plates at either

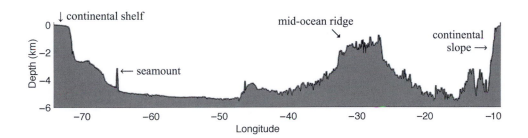

Figure 5.3 Example of ocean bathymetry: a zonal section in the Atlantic Ocean at 39° 15′ N (off shore from New Jersey to central Portugal). Figure made using the topographic database of Smith and Sandwell (1997).

side of the ridge drift away from each other. The opposite happens in the deep troughs in the Pacific Ocean (not shown), where subduction takes place and ocean floor disappears beneath the continents. Finally, the ocean basins are sprinkled with isolated seamounts, the result of ancient volcanic activity. Notice that the perception of steep slopes in Figure 5.3 is somewhat deceptive. In real dimensions, the horizontal scale (in this figure covering some 5,600 km) is much larger than the vertical scale, so slopes are predominantly mild (see also Figure 7.16). However, this does not change the fundamental importance of continental slopes as boundaries of the basin and their role in tidal propagation.

5.2 Equations of Motion

Tides propagate across the oceans, as shown in Figures 5.1 and 5.2. The curved nature of the global spherical geometry (as exemplified by the longitudinal and latitudinal circles) complicates the description of wave propagation on this scale. To keep things simple, we shall consider smaller areas, around a certain fixed latitude ϕ, where we can regard the Earth's surface as fundamentally flat. In the literature of geophysical fluid dynamics, this construct is called an f-plane, where the parameter f represents the *Coriolis parameter*, defined as

$$f = 2\Omega \sin \phi. \tag{5.1}$$

Here Ω is the Earth's angular velocity, based on the true sidereal day (see Table 3.2): $\Omega = 7.292 \times 10^{-5}$ rad/s. Notice that $f < 0$ in the southern hemisphere.

The Coriolis parameter is associated with the Coriolis force, which owes its existence to the daily rotation of the Earth on its axis. The strength of the force is proportional to f and hence vanishes at the equator. It acts on particles that move with respect to the rotating Earth. On eastward moving particles, it exerts a pull to the south in the northern hemisphere, and to the north in the southern hemisphere. On northward moving particles, it exerts a pull to the east in the northern hemisphere, and to the west in the southern hemisphere. The Coriolis force never initiates a motion but only acts as a deflecting force on an existing motion, in an energy-neutral way: its direction being perpendicular to the motion, the Coriolis force does not do work.

Figure 5.4 shows how f varies with latitude. The dimension of f invites a comparison with the frequencies of the tidal constituents, ω. For this purpose, we cast them in the same unit as f, radians per second (see Table 4.2 for their values in degrees per hour). For the four main constituents, we plot them on top of the graph of f. They all lie within the range of f. In other words, for each of these tidal constituents, there is a certain latitude at which the Coriolis parameter matches its frequency, namely $\phi = \pm \arcsin(\omega/2\Omega)$. We include a minus sign to represent the southern hemisphere as well, where f is negative. The equality $\omega = |f|$ is

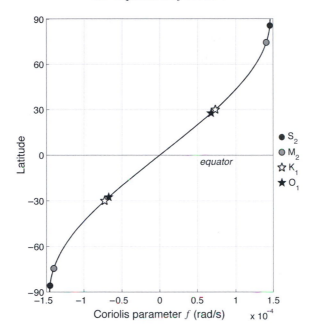

Figure 5.4 Coriolis parameter against latitude (black line), defined by (5.1), with positive (negative) values in the northern (southern) hemisphere. On top of this curve, we indicate at what latitudes the frequencies of the tidal constituents S_2, M_2, K_1, and O_1 match the Coriolis parameter, in the sense that $f = \pm \omega_{S_2}$, etc.

satisfied for diurnal frequencies at around 30° N/S, and for semidiurnal constituents at high latitudes, beyond the polar circles. The dynamical significance of this fact will become clear in Section 5.3, where we shall see that for some kinds of wave solutions, $|f|$ acts as a lower limit on the range of allowable wave frequencies; this implies that they can only exist at sufficiently low latitudes. Wave frequencies exceeding the local value of $|f|$ are called *super-inertial*, those below it, *sub-inertial*.

In this chapter, we focus on tides in the ocean basins and ignore frictional effects, as we expect them to be of minor importance for tidal waves in the deep ocean, which are characterized by weak currents (a few centimeters per second at most), as discussed in Section 1.2.3. The momentum equations on the f-plane then read

$$\frac{\partial u}{\partial t} + u\frac{\partial u}{\partial x} + v\frac{\partial u}{\partial y} + w\frac{\partial u}{\partial z} - fv = -\frac{1}{\rho}\frac{\partial p}{\partial x} \tag{5.2}$$

$$\frac{\partial v}{\partial t} + u\frac{\partial v}{\partial x} + v\frac{\partial v}{\partial y} + w\frac{\partial v}{\partial z} + fu = -\frac{1}{\rho}\frac{\partial p}{\partial y} \tag{5.3}$$

$$\frac{\partial w}{\partial t} + u\frac{\partial w}{\partial x} + v\frac{\partial w}{\partial y} + w\frac{\partial w}{\partial z} = -\frac{1}{\rho}\frac{\partial p}{\partial z} - g, \tag{5.4}$$

where t is time; x and y are the horizontal coordinates and z is the vertical coordinate (positive upward). The corresponding current velocities are u, v, and w, respectively. We consider an arbitrary but fixed latitude, i.e., we treat f as a constant. In this f-plane setting, there is no absolute geographical orientation associated with x and y; solutions remain valid if one rotates the configuration in the horizontal plane. Finally, the equations feature pressure p and density ρ; constant g is the acceleration due to gravity.

The three equations (5.2), (5.3), and (5.4) contain five unknowns (u, v, w, p, and ρ), so we need two more equations. The first one is the expression for conservation of mass

$$\frac{\partial \rho}{\partial t} + \frac{\partial}{\partial x}(\rho u) + \frac{\partial}{\partial y}(\rho v) + \frac{\partial}{\partial z}(\rho w) = 0. \tag{5.5}$$

Thermodynamic principles (discussed in Chapter 7) provide the other equation, but here we take a shortcut by simply assuming density to be constant:

$$\rho = \text{constant}. \tag{5.6}$$

In this chapter, we thus regard seawater as incompressible and uniform in density. With (5.6), the equation for conservation of mass (5.5) reduces to

$$\frac{\partial u}{\partial x} + \frac{\partial v}{\partial y} + \frac{\partial w}{\partial z} = 0. \tag{5.7}$$

5.2.1 Linear and Hydrostatic Approximations

With (5.2), (5.3), (5.4), (5.6), and (5.7) we have obtained a closed set of equations, but a couple of approximations can be made to simplify them.

The first approximation concerns the nonlinear terms, i.e., the advective terms that contain products of current velocities: for example, $u \, \partial u / \partial x$ in (5.2). We evaluate the importance of this term relative to the acceleration term $\partial u / \partial t$. We denote typical scales of the various quantities by square brackets, so $[t]$ will be the period of the tidal constituent under consideration, $[u]$ its typical amplitude of the current velocity (a few centimeters per second in the ocean), $[x]$ the horizontal scale over which u varies, etc. Thus, for the first two terms in (5.2),

$$\overbrace{\frac{\partial u}{\partial t}}^{[u]/[t]} + \overbrace{u\frac{\partial u}{\partial x}}^{[u]^2/[x]} + \cdots.$$

The ratio of the second term over the first can be expressed as

$$\frac{[u]}{[x]/[t]}. \tag{5.8}$$

For $[x]$, the wavelength is an obvious choice. From Figures 5.1 and 5.2, we can deduce empirically that its scale is of the order of several thousand kilometers (the distance between 12 successive phase lines). Filling in the typical values for $[u]$, $[t]$ and $[x]$, we find that the magnitude of (5.8) is very small, something like 10^{-3}. In other words, the nonlinear term is estimated to be much smaller than the acceleration term. One may argue that other scales for $[x]$ can be appropriate as well. After all, spatial variations in u are not only associated with the wavelength, but also with spatial variations in depth (see the bathymetry in Figure 5.3). Obstacles like large ridges and continental slopes have horizontal scales of the order of hundred kilometers, which still keeps the number (5.8) small. We can hold similar arguments for the other nonlinear terms involving derivatives to x and y. For the nonlinear terms with $w\partial/\partial z$, the case for ignoring them is similar but requires an extra step. From the continuity equation (5.7), we find that $[w]/[u] \sim [z]/[x]$; this ratio is very small because the wavelength is much larger than the water depth. Filling in this scale for $[w]$ leads to the conclusion that the terms $w\,\partial/\partial z$, too, can be neglected against the acceleration terms $\partial/\partial t$.

We thus arrive at the linear set of momentum equations

$$\frac{\partial u}{\partial t} - fv = -\frac{1}{\rho}\frac{\partial p}{\partial x} \tag{5.9}$$

$$\frac{\partial v}{\partial t} + fu = -\frac{1}{\rho}\frac{\partial p}{\partial y} \tag{5.10}$$

$$\frac{\partial w}{\partial t} = -\frac{1}{\rho}\frac{\partial p}{\partial z} - g. \tag{5.11}$$

We can make one further simplification, the *hydrostatic approximation*. The vertical acceleration in (5.11) scales as $[w]/[t]$, where the scale of the vertical velocity can be expressed as the tidal range (a_r, the vertical interval over which the water particles move up and down) divided by the time scale: $[w] = a_r/[t]$. With tidal ranges of several meters at most, this means that the acceleration term in (5.11) must be extremely small compared to the term g on the right-hand side. Hence, we can assume that the primary balance is between the two terms on the right-hand side,

$$\frac{\partial p}{\partial z} = -\rho g, \tag{5.12}$$

the *hydrostatic balance*. This equation means that the pressure at any depth level is specified by the weight of the water column above it.

The set of equations (5.6), (5.7), (5.9), (5.10), and (5.12) forms our basis for further analysis.

5.2.2 Boundary Conditions

We introduce a free ocean surface at $z = \zeta(t, x, y)$ and vertically integrate the hydrostatic balance (5.12) from level z to ζ,

$$p = \rho g(\zeta - z) + p_a, \tag{5.13}$$

where p_a is the pressure at the surface, i.e., the atmospheric pressure, which we assume to be constant. Substitution of (5.13) in (5.9) and (5.10) gives

$$\frac{\partial u}{\partial t} - fv = -g\frac{\partial \zeta}{\partial x} \tag{5.14}$$

$$\frac{\partial v}{\partial t} + fu = -g\frac{\partial \zeta}{\partial y}. \tag{5.15}$$

The right-hand sides of (5.14) and (5.15) are independent of z, so it is natural to assume that u and v, too, are independent of z. This renders the vertical integration of the continuity equation (5.7) particularly easy, as shown below.

We first define the boundary conditions at the upper and lower boundaries of the system. We introduce a flat bottom at level $z = -H$, with constant H, where we impose the boundary condition of vanishing normal flow:

$$w|_{z=-H} = 0. \tag{5.16}$$

The choice of a flat bottom may seem overly restrictive, but it allows for straightforward extensions. The point is that we can couple different domains of uniform depth to construct a basin that as a whole has a varying depth. The equations can be solved for the separate domains and then matched at their common boundaries. An example will be discussed in Section 5.7, where we connect an ocean basin with a continental shelf sea.

The boundary condition at the free surface can be derived as follows. We apply the material derivative D/Dt, defined as

$$\frac{D}{Dt} = \frac{\partial}{\partial t} + u\frac{\partial}{\partial x} + v\frac{\partial}{\partial y} + w\frac{\partial}{\partial z}, \tag{5.17}$$

to both sides of the equation for the free surface, $z = \zeta(t, x, y)$. This gives

$$w = \frac{\partial \zeta}{\partial t} + u\frac{\partial \zeta}{\partial x} + v\frac{\partial \zeta}{\partial y}, \qquad \text{at } z = \zeta. \tag{5.18}$$

We expand the left-hand side in a Taylor approximation as

$$w|_{z=\zeta} = w|_{z=0} + \zeta\frac{\partial w}{\partial z}\Big|_{z=0} + \cdots.$$

With arguments similar to those of the previous section, we can assume that all the nonlinear terms are negligible. Hence, all that remains of (5.18) at this level of approximation is

$$w|_{z=0} = \frac{\partial \zeta}{\partial t}. \tag{5.19}$$

Using (5.16) and (5.19), the vertically integrated version of (5.7), from $z = -H$ to $z = 0$, becomes

$$\frac{\partial \zeta}{\partial t} + H\left(\frac{\partial u}{\partial x} + \frac{\partial v}{\partial y}\right) = 0. \tag{5.20}$$

5.2.3 Single Governing Equation

We combine the set of three equations – (5.14), (5.15), and (5.20) – into one equation for the free surface elevation ζ. First, we take the time derivative of (5.14) and then substitute $\partial v/\partial t$ from (5.15) to eliminate v,

$$\frac{\partial^2 u}{\partial t^2} + f^2 u = -g\left(\frac{\partial^2 \zeta}{\partial t \partial x} + f\frac{\partial \zeta}{\partial y}\right). \tag{5.21}$$

Similarly, we obtain an equation for v and ζ:

$$\frac{\partial^2 v}{\partial t^2} + f^2 v = -g\left(\frac{\partial^2 \zeta}{\partial t \partial y} - f\frac{\partial \zeta}{\partial x}\right). \tag{5.22}$$

Equations (5.21) and (5.22) are known as the polarization relations; they yield u and v for given ζ. We now take $\partial/\partial y$ of (5.14), $\partial/\partial x$ of (5.15), and subtract the results:

$$\frac{\partial^2 u}{\partial t \partial y} - \frac{\partial^2 v}{\partial t \partial x} - f\left(\frac{\partial u}{\partial x} + \frac{\partial v}{\partial y}\right) = 0. \tag{5.23}$$

Or, combined with (5.20),

$$\frac{\partial}{\partial t}\left(\frac{\partial u}{\partial y} - \frac{\partial v}{\partial x} + \frac{f}{H}\zeta\right) = 0. \tag{5.24}$$

Integration in time gives

$$\frac{\partial u}{\partial y} - \frac{\partial v}{\partial x} + \frac{f}{H}\zeta = \tilde{c}(x, y), \tag{5.25}$$

where $\tilde{c}$ is a constant of integration (with respect to time), which may depend on x and y. For tidal motions, we anticipate that u, v and ζ, along with their derivatives, will all be sinusoidal in time. This makes the presence on the right-hand side of the time-independent $\tilde{c}$ incongruent, hence we must have $\tilde{c} = 0$.

Finally, we take $\partial/\partial y$ of (5.21), $\partial/\partial x$ of (5.22), subtract the results, and use (5.25) to obtain

$$\boxed{\frac{\partial^2 \zeta}{\partial t^2} - gH\left(\frac{\partial^2 \zeta}{\partial x^2} + \frac{\partial^2 \zeta}{\partial y^2}\right) + f^2 \zeta = 0.} \tag{5.26}$$

It can be shown that u and v also satisfy (5.26).

Later in this chapter, we consider boundaries in the horizontal domain, in the simple form of vertical walls. The corresponding boundary conditions are obvious: $u = 0$ at a wall placed in the y direction, and $v = 0$ at a wall placed in the x direction. As our principal equation (5.26) is in terms of ζ, we can use (5.21) and (5.22) to express the boundary conditions accordingly.

Exercise

5.2.1 Demonstrate that u satisfies the same equation as ζ in (5.26), i.e.,

$$\frac{\partial^2 u}{\partial t^2} - gH\left(\frac{\partial^2 u}{\partial x^2} + \frac{\partial^2 u}{\partial y^2}\right) + f^2 u = 0. \tag{5.27}$$

5.3 Poincaré Waves

We first consider a basin that is unbounded in both horizontal directions. The simplest way to solve (5.26) is to assume a sinusoidal plane wave solution like

$$\zeta = \zeta_0 \cos(kx + ly - \omega t), \tag{5.28}$$

with real wave numbers k and l, frequency ω, and constant amplitude ζ_0. Substitution in (5.26) gives the *dispersion relation*, i.e., the relation that connects frequency and wave numbers:

$$\omega^2 = f^2 + gH(k^2 + l^2). \tag{5.29}$$

Importantly, this expression implies that the Coriolis parameter acts as a lower bound to the range of possible wave frequencies, as we must have $\omega > |f|$. This poses a severe limitation on the latitudes at which tides may exist (so far as *this* solution is concerned); they can only be super-inertial. From Figure 5.4 we see that diurnal tides satisfy the requirement $\omega > |f|$ only in the zonal band between about 30°S and 30°N. In other words, according to (5.29), diurnal tides would not be able to exist at higher latitudes. This is plainly contradicted by the empirical evidence of Figure 5.2, which shows that they exist at all latitudes. We can thus conclude that the present plane wave solution is inadequate and that we need to find other less restrictive solutions.

We first explore a kindred wave solution for a channel of infinite length. We place the channel in the y direction, with vertical walls at $x = 0$ and $x = L$, and assume wave propagation in the along-channel direction. For the cross-channel current u, we start with the following general form

$$u = (A \sin kx + A' \cos kx) \exp i (ly - \omega t). \tag{5.30}$$

We have written the expression in complex notation and will take the real part afterwards; coefficients A and A' are (complex) constants. At the walls, u has to vanish. The condition at $x = 0$ implies $A' = 0$. To satisfy the condition at $x = L$, we must have

$$k = \frac{n\pi}{L}, \qquad \text{for } n = 1, 2, 3 \cdots, \tag{5.31}$$

with mode number n. The dispersion relation (5.29) undergoes a modification, since wavenumber k is now discretized:

$$\omega^2 = f^2 + gH \left(\left(\frac{n\pi}{L} \right)^2 + l^2 \right). \tag{5.32}$$

For given tidal frequency ω (e.g., M_2), water depth H, channel width L, latitude ϕ (whence f), and mode number n, we can determine wavenumber l from (5.32), which now also appears in discretized form:

$$l_n^2 = \frac{\omega^2 - f^2}{gH} - \left(\frac{n\pi}{L} \right)^2. \tag{5.33}$$

Notice that we can arbitrarily add different modes; the superposition automatically forms another solution because the problem is linear:

$$u = \sum_n A_n \sin(k_n x) \exp i (l_n y - \omega t), \tag{5.34}$$

where the summation involves any selection of mode numbers n.

We can readily find the corresponding fields of the along-channel current component v and surface elevation ζ. We use (5.23) to derive v from u. By substituting (5.34) in (5.23), we find that v must have the same structure as u, except that we now also need to include $\cos k_n x$ terms:

$$v = \sum_n \left[B_n \sin k_n x + C_n \cos k_n x \right] \exp i (l_n y - \omega t), \tag{5.35}$$

with complex coefficients B_n and C_n. Requiring the coefficients of each cosine and sine term to be zero after substitution in (5.23), we obtain the following set of equations for B_n and C_n:

$$\omega k_n B_n - f l_n C_n = -i f k_n A_n$$
$$f l_n B_n + \omega k_n C_n = -i \omega l_n A_n.$$

Hence, B_n and C_n are given by

$$B_n = -i\omega f \frac{k_n^2 + l_n^2}{(\omega k_n)^2 + (f l_n)^2} A_n, \qquad C_n = -i k_n l_n \frac{\omega^2 - f^2}{(\omega k_n)^2 + (f l_n)^2} A_n. \quad (5.36)$$

For convenience, we introduce the shorthand notation $B_n = \bar{B}_n A_n$ and $C_n = \bar{C}_n A_n$, where $\bar{B}_n$ and $\bar{C}_n$ represent the coefficients of A_n on the right-hand sides of (5.36). We can thus write v in (5.35) as

$$v = \sum_n A_n \left[\bar{B}_n \sin k_n x + \bar{C}_n \cos k_n x \right] \exp i(l_n y - \omega t). \quad (5.37)$$

Finally, ζ follows from (5.25), with $\tilde{c} = 0$,

$$\zeta = \frac{H}{f} \sum_n A_n \left[\bar{B}_n k_n \cos k_n x - (\bar{C}_n k_n + i l_n) \sin k_n x \right] \exp i(l_n y - \omega t). \quad (5.38)$$

Alternatively, defining the coefficient of the cosine term as $a_n = A_n \bar{B}_n k_n H / f$, and using the definitions of $\bar{B}_n$ and $\bar{C}_n$, we can write (5.38) as

$$\zeta = \sum_n a_n \left[\cos k_n x + \frac{f l_n}{\omega k_n} \sin k_n x \right] \exp i(l_n y - \omega t). \quad (5.39)$$

We call the wave structure (5.39), characterized by the dispersion relation (5.32), a *Poincaré wave*.[2] It has the character of a standing wave in the cross-channel direction while being progressive in the along-channel direction. An example is shown in Figure 5.5 (left panel), showing a snapshot of a wave propagating in the direction of the arrow. The asymmetry in the lateral direction is conspicuous: amplitudes are higher on the left-hand side, facing the direction of wave propagation (recall that the system can be arbitrarily rotated in the horizontal plane, as we are adopting the f-plane). This asymmetry is due to Coriolis effects and disappears at the equator, where $f = 0$. The present setting is for the northern hemisphere; in the southern hemisphere, amplitudes are higher on the right-hand side when facing the direction of wave propagation.

The solution shown in Figure 5.5 (left panel) involves only one mode, $n = 1$; for the parameters used here (as listed in the caption of Figure 5.5), no higher modes exist with real l. It is indeed clear from the dispersion relation (5.32) that higher mode numbers n make the restriction on the range of allowable wave frequencies more severe, as they heighten the required minimum. This is illustrated in Figure 5.5 (right panel).

We have so far assumed the wavenumber l_n to be real, but all the expressions remain valid of we replace l_n with $i l_n$. In the present setting of an infinite channel,

---

[2] In the literature, this name is sometimes also used for waves with (5.29) and (5.30) without confinement to a channel.

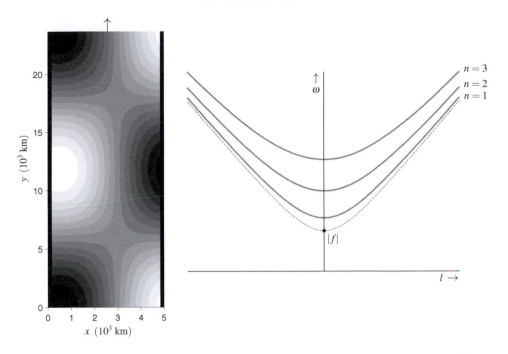

Figure 5.5 *Left:* top view of a Poincaré wave, given by the real part of (5.39), showing a snapshot at $t = 0$ of a wave propagating in the direction of the arrow. Black (white) means positive (negative) values of ζ. Parameter values are: latitude $\phi = 45°\text{N}$ ($f = 1.0 \times 10^{-4}$ rad/s), channel width $L = 5,000$ km, channel depth $H = 2$ km, wave frequency $\omega = 1.405 \times 10^{-4}$ rad/s (M_2 tidal constituent). Here $n = 1$, with $a_1 = 1$ m; higher modes do not exist for these parameters. *Right:* plot of the dispersion relation (5.32) for the first three modes and for the limiting case of large L (i.e., $k \to 0$).

this does not offer a physically meaningful solution, for the wave amplitude would grow indefinitely with y (or $-y$). However, it turns out to be a useful ingredient in a different setting with a lateral wall, as discussed in Section 5.6.

Exercise

5.3.1 Selecting one Poincaré mode n from (5.39), derive the corresponding current velocity fields u and v from the polarization relations (5.21) and (5.22). The results must of course be equivalent to mode n in (5.34) and (5.37), which forms a useful check. Selecting the first mode $n = 1$, show that v changes sign in the cross-channel direction whereas u does not. What does this imply for the sense of rotation of the current in the uv-plane?

5.4 Kelvin Waves

We now consider a wave solution that imposes no restriction at all on the range of possible wave frequencies. The setting involves (for the moment) just one wall, at $x = 0$, say, which we regard as a simple representation of a continental slope. At the wall, we must have $u = 0$, but we will in fact assume that u vanishes *everywhere*. We can then readily find the solution by returning to (5.14), (5.15), and (5.20):

$$-fv = -g\frac{\partial \zeta}{\partial x} \tag{5.40}$$

$$\frac{\partial v}{\partial t} = -g\frac{\partial \zeta}{\partial y} \tag{5.41}$$

$$\frac{\partial \zeta}{\partial t} + H\frac{\partial v}{\partial y} = 0. \tag{5.42}$$

Equation (5.40) states that there is a balance between the Coriolis force and the pressure gradient, a *geostrophic balance*, characterizing the dynamics in the lateral direction. We try a wave solution of the form

$$\zeta = F(x)\sin(ly - \omega t), \tag{5.43}$$

which describes a wave propagating along the wall, with a lateral structure $F(x)$ that we have to determine. Substitution of this expression in (5.42) gives v,

$$v = \frac{\omega}{Hl}F(x)\sin(ly - \omega t). \tag{5.44}$$

Substituting this in (5.41), we find another expression for ζ,

$$\zeta = \frac{\omega^2}{gHl^2}F(x)\sin(ly - \omega t).$$

To be consistent with (5.43), we must have

$$\omega^2 = gHl^2, \tag{5.45}$$

which is the dispersion relation. Importantly, it poses no restriction whatever on the range of possible frequencies (in contrast to the Poincaré waves discussed in the previous section). The phase speed, ω/l, is given by

$$c_0 = (gH)^{1/2}, \tag{5.46}$$

which is constant for uniform depth. The dispersion relation (5.45) and phase speed (5.46) do not contain the Coriolis parameter f. In this sense, the wave propagates as if there were no Coriolis effects. However, as we shall see, the lateral wave structure depends crucially on f. We find this structure by substituting the expressions (5.43) and (5.44) in the geostrophic balance (5.40):

$$F'(x) - \frac{f\omega}{gHl} F(x) = 0,$$

where the prime denotes the derivative to x. We can simplify the coefficient in this equation, but have to be careful with the signs. Wavenumber l is positive for propagation in the positive y direction, and negative in the opposite direction. The wave frequency will always be taken positive. So, we may write the dispersion relation (5.45) as $\omega = c_0|l|$. Hence

$$\frac{f\omega}{gHl} = \text{sgn}(l)\frac{f}{c_0},$$

since $l = |l|\,\text{sgn}(l)$. The function F thus takes the form

$$F(x) = \zeta_0 \exp\left(\text{sgn}(l)\frac{f}{c_0}x\right), \tag{5.47}$$

where ζ_0 is a constant, representing the amplitude. The solution (5.43) with (5.47) and dispersion relation (5.45) is known as a *Kelvin wave*. Notice that this dispersion relation is a special case of (5.29) with $k = \pm i f/c_0$. The exponential character of the solution in the lateral direction, seen in (5.47), is a reflection of k being imaginary.

For the wave to be exponentially decaying off the wall, the wall has to be to the right if we look in the direction of wave propagation. This applies to the northern hemisphere, as illustrated in Figure 5.6. In the southern hemisphere, where $f < 0$, the wall has to be to the left.

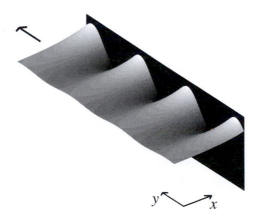

Figure 5.6 A Kelvin wave in the northern hemisphere, with the wall on its right-hand side facing the direction of wave propagation (as indicated by the arrow). The wave amplitude decays exponentially off the wall.

Since v and ζ have identical structures $\sim \sin(ly - \omega t)$, maximum current velocities coincide with maximum elevations or depressions. In terms of tidal propagation, this means that high tides are accompanied by maximum flood currents, and low tides by maximum ebb currents (the situation depicted in Figure 1.5a).

Finally, we look at some representative values for the parameters featuring in the Kelvin wave. For the ocean, with a typical depth of 4 km, the phase speed is 720 km/h. The wavelength $c_0 T$ (with wave period $T = 2\pi/\omega$), the space between successive high waters, is about 9,000 km for semidiurnal tides, and double this value for diurnal tides (cf. the spacing between co-phase lines in Figures 5.1 and 5.2). Amplitudes are highest at the wall and have a decay scale of $c_0/|f|$, the so-called Rossby radius of deformation. For mid-latitudes ($|f| \approx 10^{-4}$ rad/s), this gives a scale of 2,000 km. For example, a Kelvin wave passing at one side of the Atlantic, hardly has a signal any more at the opposite side. Furthermore, this scale provides a justification for the simple representation of a continental slope as a vertical wall. Viewed from the lateral length scale of the Kelvin wave, the width of the continental slope (typically 50–100 km) is very short and it effectively acts as a nearly vertical wall.

For coastal seas, we rather have typical depths of 40 m, which gives a phase speed of 72 km/h and a wavelength of about 900 km for semidiurnal tides. The decay scale is now 200 km. The tidal frequency plays no role in the decay scale, so all constituents have the same decay scale. The same is true for the phase speed, which is given by (5.46).

5.5 Kelvin Waves in a Channel

In the previous section, we obtained the Kelvin wave in a setting with one wall, but the solution remains valid if we add another parallel wall. After all, the lateral current velocity is already zero everywhere, and so the new boundary condition is automatically fulfilled. For the same reason, we may add another Kelvin wave travelling in the opposite direction along the newly added wall. Specifically, for a channel defined by parallel walls at $x = 0$ and $x = L$, we can write the following solution for a pair of oppositely propagating Kelvin waves:

$$\zeta = \overbrace{\zeta_0 \exp\left(-\frac{f}{c_0}x\right) \sin(ly + \omega t)}^{\text{southward}} + \overbrace{\zeta_0 \exp\left(\frac{f}{c_0}(x - L)\right) \sin(ly - \omega t)}^{\text{northward}}. \quad (5.48)$$

For convenience, we have given the two waves the same amplitude (ζ_0), but the expression is valid for any choice of amplitudes. Hereafter, the convention is that l is positive, as the correct signs have already been incorporated in (5.48) with regard to the directions of wave propagation and the decay off the wall. Here, and in the remainder of this section, we consider wave propagation in the northern hemisphere; the results can be easily adapted for the southern hemisphere by changing the sign of f and the direction of wave propagation.

For easy reference, we call the first term in (5.48) the "southward" propagating wave, and the second term, the "northward" propagating wave, but notice that the expression can be applied to any geographical orientation of the channel because the solution remains valid under a rotation in the horizontal plane.

We cast (5.48) in terms of amplitude Q and phase φ (both spatially varying) to make plots analogous to Figures 5.1 and 5.2. We thus seek to find functions Q and φ such that

$$\zeta = Q \cos(\omega t - \varphi). \tag{5.49}$$

With (A.2), we can write (5.49) as

$$\zeta = Q \cos\varphi \cos\omega t + Q \sin\varphi \sin\omega t. \tag{5.50}$$

On the other hand, using (A.1) we can write (5.48) as

$$\zeta = \overbrace{(E_m + E_p)\sin ly}^{Q_c} \cos\omega t + \overbrace{(E_m - E_p)\cos ly}^{Q_s} \sin\omega t, \tag{5.51}$$

where we introduced the short–hand notation Q_c and Q_s for the coefficients, and

$$E_m = \zeta_0 \exp(-fx/c_0), \qquad E_p = \zeta_0 \exp(f(x-L)/c_0). \tag{5.52}$$

Comparing (5.50) and (5.51), we obtain the expressions for Q and φ:

$$Q = (Q_c^2 + Q_s^2)^{1/2}, \qquad \cos\varphi = Q_c/Q, \qquad \sin\varphi = Q_s/Q. \tag{5.53}$$

The pair of Kelvin waves (5.48) can thus be represented in terms of amplitude $Q(x, y)$ and phase $\varphi(x, y)$.

For the along-channel current v, we obtain an expression by combining (5.42) and (5.51):

$$v = \frac{\omega}{Hl}(E_p - E_m)\sin ly \cos\omega t - \frac{\omega}{Hl}(E_p + E_m)\cos ly \sin\omega t. \tag{5.54}$$

The amplitude and phase of v can now be obtained in an analogous way as for ζ.

An example of a pair of oppositely propagating Kelvin waves is shown in Figure 5.7. The amplitude and phase are plotted for sea surface elevation ζ and along-channel current velocity v. The combination of oppositely propagating Kelvin waves creates *amphidromic points*, where the amplitude vanishes and the co-phase lines go around, which we encountered earlier in Figures 5.1 and 5.2. Notice that the amphidromic points of sea surface elevation ζ and current velocity

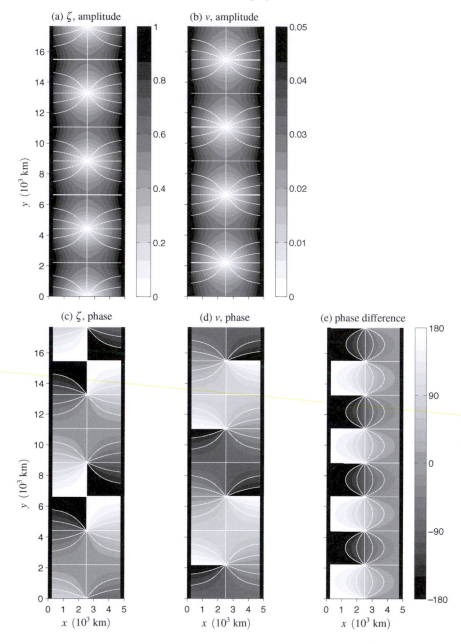

Figure 5.7 Combination of southward and northward propagating Kelvin waves in a channel (top view). Amplitude (in meters) and phase (in degrees) are plotted for sea surface elevation ζ in (a) and (c). Similarly for the along-slope current velocity v in (b) and (d), with the amplitude in m/s. The gray bar in (e) applies also to (c) and (d). In (e) the phase difference between surface elevation and current velocity is shown, i.e., (c) minus (d). The parameters are: channel width $L = 5 \times 10^3$ km, water depth $H = 4$ km, latitude $\phi = 45°$N, wave amplitude $\zeta_0 = 1$ m, tidal frequency $\omega = 1.405 \times 10^{-4}$ rad/s (M$_2$).

v lie at *different* positions. Where the sea surface is motionless, there is still a tidal current (and vice versa). Amplitudes are highest at the walls. The definition of the phase in expression (5.49) implies that phase propagation is towards higher values of φ. Hence, phases propagate northward in the eastern half of the channel, and southward in the western half. (Notice that in the gray scale of the phases, black represents the same phase as white.) Also plotted is the phase difference between ζ and v (Figure 5.7e). At the eastern wall, the phase difference is zero, implying that ζ and v are in phase: high tides coincide with maximum currents in the direction of wave propagation, low tides with maximum currents in the opposite direction. This is also the case at the other wall, but now with a phase difference of $\pm 180°$ as maximum currents in the direction of wave propagation point in the negative y direction. At the walls, the behavior is thus like in a single Kelvin wave (cf. (5.43) and (5.44)). Towards the center of the channel, however, the phase difference goes through a range of values, including $\pm 90°$, where maximum elevations coincide with slack waters. The three scenarios sketched in Figure 1.5 all come into play already in this relatively simple setting.

Exercise

5.5.1 Without using any formula, argue what happens to the amphidromic points in Figure 5.7 when the amplitudes of the oppositely travelling Kelvin waves are unequal.

5.6 The Taylor Problem

So far, we have considered basins shaped by infinitely long walls or channels. As a step towards a more realistic setting, we now add a wall in the lateral direction, at $y = 0$, say. This turns the channel into a *semi-enclosed basin*. We examine the propagation of tides in such a basin; this is called the *Taylor problem*.

The starting point is the solution from the previous section, with two Kelvin waves running in opposite directions through a channel. In the solution (5.54), now denoted by v_k, we add an arbitrary phase shift Λ in the y direction. In complex notation, the expression then becomes

$$v_k = \frac{\omega}{Hl}\Big[(E_p - E_m)\sin(ly + \Lambda) - i\,(E_p + E_m)\cos(ly + \Lambda)\Big]\exp(-i\omega t). \quad (5.55)$$

The real part is implied. It turns out that we need the additional parameter Λ to solve the problem. In the setting of a semi-enclosed basin, we can interpret (5.55) as a combination of incident and reflected Kelvin waves. However, returning to Figure 5.7b, we see that there is no lateral transect at which the along-channel current vanishes uniformly. In other words, no matter how we shift the wave pattern by adapting Λ, (5.55) can never satisfy the boundary condition at the lateral wall. Thus, (5.55) by itself does not provide a solution for the semi-enclosed basin. For this reason, we seek an *additional* solution that solves our problem at the lateral wall, while not affecting the Kelvin waves far away from it.

With a simple modification, we can obtain the required additional solution from the Poincaré waves considered in Section 5.3. With (5.37), we have an expression for progressive Poincaré waves in a channel. We now turn this into a solution that is evanescent in the up-channel direction by replacing wavenumber l_n with il'_n (l'_n real and positive):

$$v_p = \sum_n A_n \exp(-l'_n y) \left[\bar{B}_n \sin k_n x + \bar{C}_n \cos k_n x \right] \exp(-i\omega t). \tag{5.56}$$

Correspondingly, $\bar{B}_n$ and $\bar{C}_n$, the coefficients of A_n in (5.36), now read

$$\bar{B}_n = -i\omega f \frac{k_n^2 - (l'_n)^2}{(\omega k_n)^2 - (f l'_n)^2}, \qquad \bar{C}_n = k_n l'_n \frac{\omega^2 - f^2}{(\omega k_n)^2 - (f l'_n)^2}. \tag{5.57}$$

Furthermore, the dispersion relation (5.33) becomes

$$(l'_n)^2 = \frac{f^2 - \omega^2}{gH} + \left(\frac{n\pi}{L} \right)^2. \tag{5.58}$$

For sub-inertial tidal frequencies ($\omega < |f|$), l'_n is real for all mode numbers. For super-inertial tidal frequencies ($\omega > |f|$), however, l'_n may turn out to be imaginary for the lowest mode numbers, in which case (5.56) would contain progressive modes in the up-channel direction (which brings us back to Section 5.3). We restrict our analysis to the case in which l'_n is real for *all* mode numbers, which means that we suppose the channel to be sufficiently narrow in the super-inertial case, in the sense that $L < \pi c_0/(\omega^2 - f^2)^{1/2}$, with c_0 defined by (5.46). All modes are then evanescent in the up-channel direction.

By itself, (5.56) does not provide a solution for the semi-enclosed basin, since v_p does not vanish at $y = 0$. However, by combining the pair of Kelvin waves (5.55) and the trapped Poincaré waves (5.56), the problem can be solved. Notice that the amplitudes of the incident and reflected Kelvin are here taken equal, which means that their reflection is assumed to be perfect. This is in line with the assumption that all Poincaré modes are trapped. If one would include progressive outgoing Poincaré waves, or frictional effects, then allowance should be made for a weakened reflected Kelvin wave.

We now impose the central requirement that v_p, from the trapped Poincaré waves, plus v_k, from the pair of Kelvin waves, *together* vanish at the lateral wall:

$$v_p|_{y=0} + v_k|_{y=0} = 0. \tag{5.59}$$

Substitution of (5.55) and (5.56) gives

$$\sum_{n=1}^{\infty} A_n \left[\bar{B}_n \sin k_n x + \bar{C}_n \cos k_n x \right] + \frac{\omega}{Hl} \left[(E_p - E_m) \sin \Lambda - i (E_p + E_m) \cos \Lambda \right] = 0.$$

$$(5.60)$$

This equation is to be satisfied for all $0 \leq x \leq L$. Recall that E_m and E_p, defined in (5.52), are functions of x. For given parameters ω, f, L, and H, we need to find the coefficients A_n and the phase Λ.

A_n and Λ can be solved numerically by requiring that (5.60) is satisfied at N selected positions, $0 < x_1 < x_2 < \cdots < x_N < L$. This method is known as the *collocation method*. The points $x_1, \cdots, x_N$ are called collocation points. We first bring the second term in (5.60), the Kelvin-wave part, to the right-hand side:

$$\sum_{n=1}^{\infty} \left[\bar{B}_n \sin k_n x + \bar{C}_n \cos k_n x \right] A_n = -\frac{\omega}{Hl} \left[(E_p - E_m) \sin \Lambda - i (E_p + E_m) \cos \Lambda \right].$$

$$(5.61)$$

Let us assign, for the moment, an arbitrary value to Λ. Then, at each collocation point, the right-hand side of (5.61) can be evaluated. The part in square brackets on the left-hand side can also be evaluated. Doing this for all the collocation points, and truncating the series at $n = N$, we obtain a set of N equations for $A_1, \cdots, A_N$. This linear system can be written in matrix form and solved by matrix inversion. It is then guaranteed that the boundary condition (5.60) is satisfied *at* the collocation points $x_1, \cdots, x_N$.

At this point, we are left with two problems. First, there is as yet no guarantee that the boundary condition will be satisfied at points lying *in-between* the collocation points. Or, put differently, we have no guarantee that the process converges for increasing N. Second, we still have to find a criterion from which we can determine Λ (which we have so far regarded as arbitrary). It turns out that the two problems are linked; they will be solved at once.

We can readily test how well the solution performs for intermediate points by subdividing the interval $(0, L)$ into a finer grid and evaluating the left-hand side of (5.60) at those grid points. The root mean square of the resulting values can be regarded as a measure of the error of the solution, indicating how well it behaves in-between the collocation points: the closer to zero, the more accurate the solution.

More generally, we can carry out this procedure for the entire range of possible values of Λ, i.e., $0 \leq \Lambda \leq 2\pi$. An example is shown in Figure 5.8. The error drops to zero for two specific values of Λ, which are π apart. It is for these values that the method converges to a uniformly valid solution in the interval $(0, L)$. The periodicity in Λ reflects the periodicity of the Kelvin wave solution in the channel, and it is immaterial which of the two optimal values for Λ is selected.

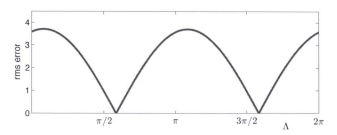

Figure 5.8 The root mean square error indicating which values of Λ provide the closest approximation to (5.60). The parameters are: channel width $L = 465$ km, water depth $H = 75$ m, latitude $\phi = 53°$N, tidal frequency $\omega = 1.405 \times 10^{-4}$ rad/s (M_2). The error drops to zero for $\Lambda = 1.834$ ($+\pi$). This result is based on a truncation at $N = 100$, with 10^3 uniformly spaced test points.

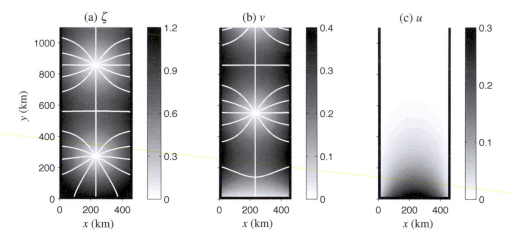

Figure 5.9 Solution of the Taylor problem in a semi-enclosed basin for the semidiurnal constituent M_2. Elevation ζ is in meters and current components u and v in m/s. Parameters H, L, ϕ, and N are as in Figure 5.8; furthermore, $\Lambda = 1.834$ and $\zeta_0 = 1$ m.

For the selected Λ, we calculate the corresponding coefficients A_n from (5.61). The fields ζ, v and u are now obtained as follows. For v, the solution is formed by the superposition of the real parts of (5.55) and (5.56). Component u was uniformly zero for the two-Kelvin wave solution, but now comes into play via (5.34), with $l_n = i l_n'$. Finally, with u and v known, ζ is obtained from (5.25). An example is shown in Figure 5.9. It is clear from Figure 5.9b that v vanishes at the lateral boundary, as required. The component u is present in the vicinity of the lateral wall, but decays in the up-channel direction (Figure 5.9c).

In reality, we are not dealing with just one tidal constituent but with a large number of them, all at once. So, it is interesting to examine the pattern of another constituent, e.g., the diurnal K_1. The resulting solution is shown in Figure 5.10. The

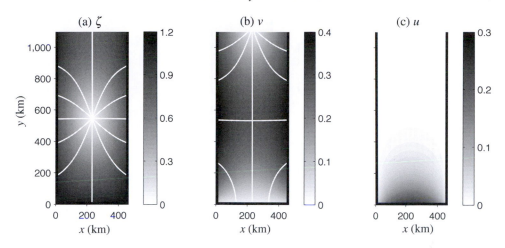

Figure 5.10 As in Figure 5.9, but now for the diurnal constituent K_1. Here, $\Lambda = 1.681$.

wave pattern is stretched in the up-channel direction (notice that the basin size was left unchanged), shifting the amphidromic points as well. The amphidromic point in Figure 5.10a is located at nearly the same place as the amphidromic point of v for the semidiurnal constituent M_2 (Figure 5.9b). As a result, currents are almost purely diurnal near that point, while sea surface elevations are semidiurnal.

The horizontal currents u and v, shown in Figures 5.9 and 5.9, can alternatively be plotted in the uv-plane, which illustrates how the currents change through the tidal cycle (Figure 5.11). The initial state ($t = 0$) is marked by a dot. In the course of a tidal period, the horizontal current components trace an ellipse in the uv-plane. We stop shortly before the end of the period; the gap allows us to deduce in what sense the ellipse is traversed. The sense of rotation is counterclockwise. Far away from the lateral boundary, the ellipses become rectilinear in the along-channel direction, as the solution is effectively described by the two Kelvin waves. Nearer to the lateral boundary, the major axis of the ellipses turns from the v to the u direction, and near the lateral boundary the ellipses become rectilinear again, but now aligned in the lateral direction.

The parameters chosen in Figures 5.9 and 5.10 are broadly representative for the North Sea basin, viewed as a tilted rectangular basin. In Figure 5.12 we show a map of M_2 in the North Sea. A direct comparison is not possible, because of the limitations in the Taylor solution as constructed in this section. First, it does not take into account the entrance of a tidal wave from the Channel (no such opening exists in Figure 5.9), which interferes with the tidal wave entering from the Atlantic Ocean along the Scottish coast. Second, the large variations in depth present the North Sea are not accounted for in the model. A third factor is the neglect of friction, which would displace the amphidromic points to the east: by the time the tidal wave reaches the eastern part of the basin, it has already lost part of its energy.

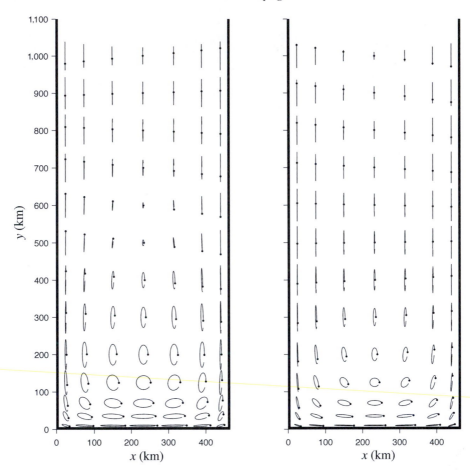

Figure 5.11 Current ellipses for M_2 (left) and K_1 (right).

5.7 Modified Kelvin Waves in the Presence of a Shelf

We have so far considered basins of uniform depth. This setup is attractive for its simplicity but misses a prominent aspect of the real ocean: the complex bathymetry and associated variations in depth (Figure 5.3). We now introduce one such element: the transition from the deep ocean to the shallow continental shelf sea. We shall still regard the continental slope as a vertical wall, but now with a narrow opening at the top, as depicted in Figure 5.13. The shelf sea has a width L. We assume that the deep ocean and continental shelf sea each have a uniform depth, H and H_s, respectively. Thus, at either side of the continental slope the governing equation (5.26) applies, but with H replaced by H_s over the shelf. In this configuration the dynamics will still be partly reminiscent of a Kelvin wave, but with fundamental modifications.

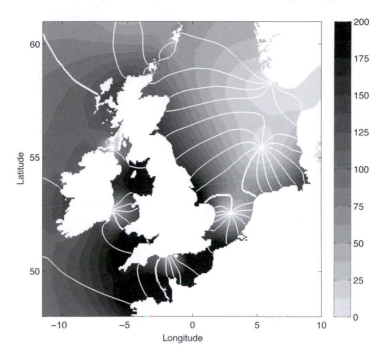

Figure 5.12 Map of the M_2 constituent in the North Sea, with amplitude in centimeters. Co-phase lines (i.e., lines connecting points of equal phase) are shown in white. Lags between neighboring lines represent a phase difference of 1/12 of the M_2 period (i.e., slightly more than one solar hour). Figure generated using Aviso+ products, courtesy of LEGOS/Noveltis/CNES/CLS.

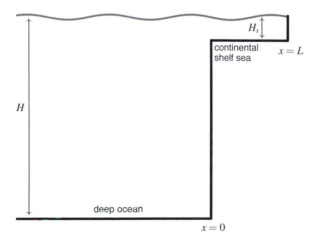

Figure 5.13 Schematic representation of the deep ocean and adjacent continental shelf.

We assume that the tide propagates along the continental slope as $\sim \exp i(ly - \omega t)$, so that we can write

$$\zeta = \hat{\zeta}(x)\exp i(ly - \omega t), \quad u = \hat{u}(x)\exp i(ly - \omega t), \quad v = \hat{v}(x)\exp i(ly - \omega t),$$

where the functions $\hat{\zeta}(x)$, $\hat{u}(x)$ and $\hat{v}(x)$ are to be determined. In each expression, the real part is implied. The wavenumber l can be found from a dispersion relation, discussed below. It will no longer be given by the simple expression of the classical Kelvin wave, but undergoes a modification due to the presence of the shelf. Furthermore, unlike in the Kelvin-wave solution, we do not assume the current component u to be zero. From (5.26) we obtain the following equation for $\hat{\zeta}(x)$:

$$\hat{\zeta}'' - \left(l^2 - \frac{\omega^2 - f^2}{gH_{(s)}}\right)\hat{\zeta} = 0, \tag{5.62}$$

where the prime denotes the derivative to x. The current velocity functions $\hat{u}(x)$ and $\hat{v}(x)$ can be expressed in terms of $\hat{\zeta}$ and its derivative, using (5.21) and (5.22),

$$\hat{u} = \frac{ig}{\omega^2 - f^2}(fl\hat{\zeta} - \omega\hat{\zeta}') \tag{5.63}$$

$$\hat{v} = \frac{g}{\omega^2 - f^2}(\omega l\hat{\zeta} - f\hat{\zeta}'). \tag{5.64}$$

We assume that the factor in brackets in (5.62) is positive over the deep ocean and introduce k as

$$k = \left(l^2 - \frac{\omega^2 - f^2}{gH}\right)^{1/2}. \tag{5.65}$$

For the shelf sea, we consider the two possible cases

$$\text{Case I:} \quad l^2 - \frac{\omega^2 - f^2}{gH_s} < 0 \quad \text{and} \quad k_s = \left(\frac{\omega^2 - f^2}{gH_s} - l^2\right)^{1/2} \tag{5.66}$$

$$\text{Case II:} \quad l^2 - \frac{\omega^2 - f^2}{gH_s} > 0 \quad \text{and} \quad k_s = \left(l^2 - \frac{\omega^2 - f^2}{gH_s}\right)^{1/2}. \tag{5.67}$$

Thus, in what follows, k and k_s are always real.

There are four boundary conditions to be imposed. At the transition between ocean and shelf $(x = 0)$, both the surface elevation and the cross-slope transport must be continuous:

$$\hat{\zeta}|_{x=0^-} = \hat{\zeta}|_{x=0^+} \tag{5.68}$$

$$H\hat{u}|_{x=0^-} = H_s\hat{u}|_{x=0^+}. \tag{5.69}$$

Furthermore, we require that the solution in the deep ocean decays off the wall (like a Kelvin wave), while the solution on the shelf has a vanishing normal current at its outer boundary:

$$\hat{\zeta} \to 0 \quad \text{for } x \to -\infty \tag{5.70}$$

$$\hat{u} = 0 \quad \text{at } x = L. \tag{5.71}$$

5.7.1 Case I: Sinusoidal Solution on Shelf

With k defined by (5.65) and k_s by (5.66), the solution of (5.62) is exponential over the deep ocean and sinusoidal over the shelf:

$$\hat{\zeta} = \begin{cases} \zeta_0 \exp kx & x < 0 \\ \zeta_0 (\cos k_s x + \gamma \sin k_s x) & x > 0. \end{cases} \tag{5.72}$$

Here ζ_0 is an arbitrary positive constant; γ is determined below. Over the deep ocean, we naturally selected the exponential solution that decays off the wall, in accordance with (5.70). Moreover, we have arranged the solution such that $\hat{\zeta}$ already satisfies (5.68).

In the present setting, the factor in brackets in (5.62) is positive over the deep ocean and negative over the shelf, so we have the combined inequalities

$$\frac{\omega^2 - f^2}{gH} < l^2 < \frac{\omega^2 - f^2}{gH_s}. \tag{5.73}$$

This acts as a constraint on wavenumber l. Moreover, since $H_s < H$, the inequalities can hold only if $\omega > |f|$, i.e. for super-inertial frequencies. This poses a severe restriction on the zones where tides of the form (5.72) can exist. For diurnal constituents, in particular, all latitudes poleward of about 30° N/S are excluded (see Figure 5.4). The solution (5.72) thus exhibits similar constraints as the Poincaré waves of Section 5.3. As we shall see below, it may or may not be possible to satisfy (5.73), depending on the parameters H, H_s, L, f, and ω.

The boundary condition (5.69) implies, via (5.63) and (5.72),

$$H(fl - \omega k) = H_s(fl - \omega k_s \gamma),$$

so that

$$\gamma = \frac{\omega k H - fl(H - H_s)}{\omega k_s H_s}. \tag{5.74}$$

Finally, we impose (5.71), which implies, again via (5.63) and (5.72),

$$\tan k_s L + \frac{fl - \omega k_s \gamma}{fl\gamma + \omega k_s} = 0. \tag{5.75}$$

This is the dispersion relation. For given constants H, H_s, L, f, and ω, we use the definitions of k in (5.65), k_s in (5.66), and γ in (5.74) to obtain with (5.75) an expression for wavenumber l. The resulting equation is transcendental and cannot be solved by analytical means. However, it can easily be solved numerically, for instance by running through the range of possible values of l and inspect for which value – if indeed for any – the left-hand side of (5.75) vanishes. Notice that f and l are positive in the northern hemisphere, and negative in the southern hemisphere; unless stated otherwise, we shall restrict ourselves to the former case to simplify the presentation.

An example of the resulting solution is shown in Figure 5.14. The presence of the shelf has a number of consequences. First, the tidal wave is shortened (i.e., l

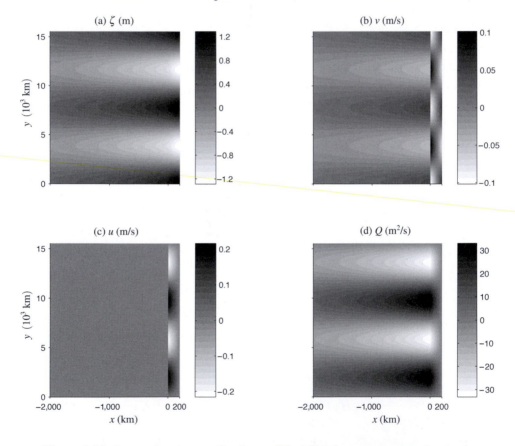

Figure 5.14 A snapshot (at $t = 0$) of a modified Kelvin wave propagating in the positive y direction (top view). In each panel, the deep ocean is on the left ($x < 0$), and the shelf on the right ($x > 0$). Panels represent (a) surface elevation ζ, (b) along-slope current component v, (c) cross-slope current component u, and the instantaneous cross-slope flux Q. Amplitude $\zeta_0 = 1$ m. The other parameters are: ocean depth $H = 4$ km, shelf depth $H_s = 150$ m, shelf width $L = 200$ km, latitude $\phi = 45°$N, and tidal frequency $\omega = 1.405 \times 10^{-4}$ rad/s (M$_2$). The dispersion relation (5.75) gives the corresponding along-slope wavelength: $2\pi/l = 7787$ km.

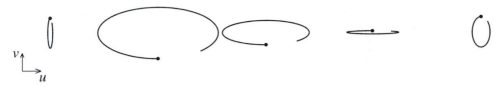

Figure 5.15 Current ellipses, tracing u and v in time. In each case, the center of the ellipse represent the origin $u = v = 0$. The starting point is indicated by a dot; the subsequent evolution during the tidal period is left incomplete to indicate the sense of rotation. The selected locations are, from left to right: $x = -25$ km (deep ocean), $x = 25$, 75, 125, and 175 km (shelf). Parameters are as in Figure 5.14.

is larger) compared with the classical Kelvin wave for the same ocean depth and tidal frequency, in this case by as much as a thousand kilometers. This implies a reduction of the phase speed ω/l. The deep ocean no longer is the only determining factor in the phase speed (as it was for the Kelvin wave with $c_0 = (gH)^{1/2}$), but the shallow shelf puts, so to speak, a brake on it. Second, the phase of the along-slope current reverses across the slope (Figure 5.14b). Third, in a fundamental departure from the classical Kelvin wave, there is a now a *cross*-slope current, which is clearly visible over the shelf (Figure 5.14c). In fact, it stretches out to the ocean side, but is much weaker there. An alternative way to visualize this component is by looking at the cross-slope transport Q, i.e., u times local water depth (Figure 5.14d). In accordance with the boundary condition (5.69) that we imposed, the transport is continuous across the slope, where it also takes its largest value.

Returning to the horizontal current components, we plot their evolution in time for a few selected locations (Figure 5.15). The initial state ($t = 0$) is marked by a dot. In the course of a tidal period, the horizontal current components trace an ellipse in the uv-plane. We stop shortly before the end of the period; the gap indicates in what sense the ellipse is traversed (i.e., the polarization). In the deep ocean (outer left), the ellipse is almost rectilinear and dominated by the v component; the sense of rotation is counterclockwise. At the other side of the slope, component u gains prominence and the sense of rotation reverses. Farther onto the shelf, however, the ellipse becomes elongated again, but now in the u direction, and again changes its sense of rotation. Finally, as the boundary at $x = L$ is approached, the along-slope v becomes again dominant.

Figure 5.14 illustrates just one example from the parameter space spanned by H, H_s, L, f, and ω. At low latitudes, we can obtain solutions for (super-inertial) diurnal tides in the same way; the result is qualitatively similar to Figures 5.14 and 5.15 and is therefore not shown. Quantitatively, the main difference is an approximate doubling of the wavelength.

We explore the parameter space further by looking at the limiting values of slope length L. The lower limit is of course $L \to 0$ (no shelf at all). Then (5.75) implies

$fl - \omega k_s \gamma = 0$, hence with (5.74): $\omega k = fl$, which naturally leads us back to the classical Kelvin wave of Section 5.4.

The shelf width L has an upper limit as well. As L increases, so does the wavenumber l. This can be verified numerically, but it also makes sense intuitively. After all, with larger L, one expects the shelf to have a more notable fingerprint on the phase speed ω/l; its shallowness moderates the phase speed, hence l should increase. However, l cannot grow indefinitely, because of the upper bound in (5.73):

$$l_* = \left(\frac{\omega^2 - f^2}{g H_s} \right)^{1/2}.$$

As $l \to l_*$, $k_s \to 0$, according to its definition (5.66). Parameter γ in (5.74) tends to infinity, but the product $k_s \gamma$ approaches the finite constant

$$\gamma_* = \frac{\omega k_* H - fl_*(H - H_s)}{\omega H_s}.$$

Here k_* is given by (5.65), with $l = l_*$. The constant γ_* features in the solution (5.72) on the shelf, which for $k_s \to 0$ becomes $\zeta_0(1 + \gamma_* x)$, attaining its maximum at the coast ($x = L$).

Finally, the value of L for which this limiting case occurs can be found from (5.75). As $\tan k_s L \to k_s L$, the maximum shelf width is given by

$$L_* = \frac{\omega \gamma_* - fl_*}{fl_* \gamma_*}. \tag{5.76}$$

Keeping all other parameters in Figure 5.14 the same, we find $L_* = 492$ km. To allow L to become still larger, we need to turn to the solution of Case II.

5.7.2 Case II: Exponential Solution on Shelf

With k_s given by (5.67), the solution of (5.62) becomes exponential over the shelf:

$$\hat{\zeta} = \begin{cases} \zeta_0 \exp kx & x < 0 \\ \zeta_0(\cosh k_s x + \gamma \sinh k_s x) & x > 0. \end{cases} \tag{5.77}$$

On the shelf, we have (via the hyperbolic functions) terms of both signs in the exponent, i.e., $\exp(\pm k_s x)$. The solution features a parameter γ, for which we use the same symbol as in the previous case, for it turns out to be again given by (5.74). In (5.77), the boundary conditions (5.68) and (5.70) are already fulfilled.

In this setting, the factor in brackets in (5.62) is positive over the deep ocean and the shelf. If $\omega < |f|$, there is no restriction on l; if $\omega > |f|$, we have a lower bound for l defined by

$$l^2 > \frac{\omega^2 - f^2}{g H_s} = l_*^2.$$

Case II extends the range of l precisely from the point where Case I reached its upper limit (l_*). Unlike in Case I, there is now no a priori constraint on the wave frequency: under the form (5.77), diurnal and semidiurnal tides can in principle exist at all latitudes (this depends however on the choice of H, H_s, and L, as we shall see below).

The boundary condition (5.69) again leads to the expression for γ stated in (5.74). The boundary condition (5.71) yields the dispersion relation

$$\tanh k_s L + \frac{fl - \omega k_s \gamma}{fl\gamma - \omega k_s} = 0. \tag{5.78}$$

For water depths H and H_s, shelf width L and latitude ϕ as in Figure 5.14, there is now a *diurnal* tide, e.g., K_1 (at this latitude sub-inertial). The solution is qualitatively similar to the one of Figures 5.14 and 5.15 (again with a change in wavelength as noted under Case I), and is therefore not shown. For the same H, H_s, and ϕ there is also semidiurnal tide, but only if we take the shelf width L sufficiently large. The maximum L in Case I, given by (5.76), now serves as the minimum value.

The transition between Case I and Case II is seamless. The transition is defined by $L \to L_*$, and accordingly, $l \to l_*$ and $k_s \to 0$. For both cases, the solution approaches $\zeta = \zeta_0(1 + \gamma_* x)$ in this limit. The transition from sinusoidal (Case I) to exponential (Case II) behavior thus happens through the approximation for small $k_s x$, where they find common ground.

A synthesis of Cases I and II is shown in Figure 5.16, where the transition between the cases is indicated by the curve L_* as a function of latitude. For any given shelf width, Case I applies at sufficiently low latitudes, and Case II at sufficiently high latitudes. In reality, the shelf width is at most a few hundreds of kilometers large, so the far left end of the panels in Figure 5.16 is the most relevant part. The transition between the cases then approximately coincides with the transition from sub- to super-inertial.

Even though very long shelves are not relevant from an oceanographic point of view, they nevertheless offer a theoretically interesting limiting case. A comparison between Figure 5.16a and 5.16b suggests that the tail end for large L is identical in both cases. This can be verified from (5.76). At low latitudes, in the sense that $f \ll \omega$, the asymptotic behavior becomes $L_* \sim (gH_s)^{1/2}/f$, which is independent of the wave frequency ω.

For very large shelf width L, Case I effectively shrinks to equatorial latitudes and so Case II becomes predominant. As $k_s L \to \infty$, we have $\tanh kL \to 1$ in the dispersion relation (5.78), which thus reduces to

$$(fl - \omega k_s)(\gamma + 1) = 0.$$

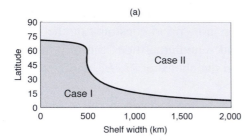
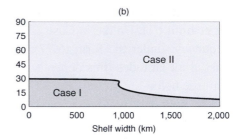

Figure 5.16 Domains of Cases I and II for ocean depth $H = 4$ km and shelf depth $H_s = 150$. The curve depicts L_*, given by (5.76), as a function of latitude, marking the transition between the cases. In (a) for the semidiurnal tide M_2; in (b) for the diurnal K_1.

There are two ways to satisfy this dispersion relation. The first is $\omega k_s = fl$, which characterizes a Kelvin wave. Far onto the shelf, the term $\exp k_s x$ becomes very large and dominates the whole signal. This represents a Kelvin wave propagating on the *shelf* at a phase speed specified by the shelf depth H_s. The coast at $x = L$ figures as the wall. By comparison, the signal becomes negligibly small towards the ocean. So, we have the opposite situation from where we started, which was a Kelvin wave in the ocean modified by a shelf.

The other solution is $\gamma = -1$. Then (5.77) reduces to

$$\hat{\zeta} = \begin{cases} \zeta_0 \exp(kx) & x < 0 \\ \zeta_0 \exp(-k_s x) & x > 0. \end{cases} \tag{5.79}$$

The amplitude takes its maximum over the slope and decays in both directions. This solution is known as a *double Kelvin wave* (not to be confused with the pair of Kelvin waves in a channel, as in Section 5.5). The remote wall at $x = L$ now has become a by–stander and we might as well regard the situation as one of a semi-infinite shelf. Wavenumber l follows from (5.74) with $\gamma = -1$:

$$(\omega k - fl)H + (\omega k_s + fl)H_s = 0. \tag{5.80}$$

After some algebraic manipulation (using the definitions of k and k_s), one can derive from (5.80) a quadratic equation for l^2:

$$\overbrace{g^2[(H+H_s)^2\omega^2 - (H-H_s)^2 f^2]}^{a}l^4 \overbrace{-2g\omega^2(H+H_s)(\omega^2-f^2)}^{b}l^2 \overbrace{+\omega^4(\omega^2-f^2)}^{c} = 0. \tag{5.81}$$

For the discriminant to be positive, i.e. $b^2 - 4ac > 0$, wave frequencies have to be sub-inertial, $\omega < |f|$. This is a necessary condition for having a solution, but not a sufficient one. In the transition from (5.80) to (5.81) squares were taken, hence solutions of (5.80) must satisfy (5.81), but not necessarily the other way round. From numerical inspection it follows that the relevant root of (5.81) is the following

$$l = \text{sgn}(f) \left(\frac{-b - (b^2 - 4ac)^{1/2}}{2a} \right)^{1/2}, \qquad (5.82)$$

where we included the case of the southern hemisphere ($f < 0$). For sub-inertial waves (as required), b is positive, hence the numerator is negative. This implies an additional constraint, $a < 0$, which further limits the frequency range; ω has to be sufficiently far into the sub-inertial regime:

$$\omega < |f| \frac{H - H_s}{H + H_s}. \qquad (5.83)$$

An important general conclusion from this section is that the presence of a continental shelf profoundly changes the character of the Kelvin wave, in that it now involves a *cross-slope* current component. This has literally far-reaching implications. A tidal current over a slope implies a vertical oscillation. In this chapter, we assumed the density to be constant, but in the real ocean, a vertical stratification in density exists. A periodic movement up and down the slope brings the levels of equal density (*isopycnals*) into motion, from which internal waves radiate at the tidal frequency. These *internal tides* propagate away from their source into the deep ocean and onto the shelf. Globally, a significant amount of energy is transferred from surface tides to internal tides, which, in turn, contribute to deep ocean mixing (as mentioned in Section 1.2.3). Internal tides are further discussed in Chapter 7.

Exercise

5.7.1 Derive the expressions for the current velocities u and v for the double Kelvin wave. Examine the sense of rotation in the uv-plane (in the same way as in Figure 5.15), distinguishing the shelf and the deep ocean. How do these characteristics change if you take the case of the southern hemisphere?

Further Reading

Based on global tidal models (Lyard and Le Provost, 1997), or in combination with data from satellite altimetry (Egbert and Ray, 2001), global maps for the M_2 constituent have been constructed that reveal where the tide-generating force injects energy into the system; the dominant areas turn out to be the South Atlantic and western Indian Oceans.

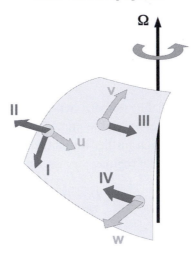

Figure 5.17 The working of the four Coriolis terms, illustrated for the northern hemisphere. For an eastward velocity u, the Coriolis force has a southward (I) and an upward (II) component; for a northward velocity v, an eastward component (III), and for an upward velocity w, a westward component (IV). In the southern hemisphere, I is northward and III westward, while II and IV stay the same. Components II and IV are neglected under the Traditional Approximation. After Gerkema and Gostiaux (2012).

The starting point in Section 5.2 is the set of equations familiar from continuum mechanics with the additional feature of the Coriolis force, which makes them applicable to geophysical fluid dynamics. The fundamental postulates underlying the momentum equations and the equations for conservation of mass and energy are treated in, for example, Serrin (1959) and Lin and Segel (1974). A derivation of the Coriolis force by a transformation of coordinates to a rotating system can be found in textbooks on classical mechanics, or, e.g., in Vallis (2006). For a general orientation of a rotating frame one finds six Coriolis terms, but the geophysical setting is special because the rotation axis lies in a plane spanned by a meridional great circle. This reduces the number of terms to four. For the steps and approximations leading from a spherical geometry to the Cartesian local f-plane equations, see LeBlond and Mysak (1978).

The working of the four Coriolis terms is depicted in Figure 5.17. They fall into two categories. First, terms I and III, which feature in (5.2) and (5.3). They are proportional to the sine of latitude and play out exclusively in the horizontal plane. Second, terms II and IV, which were tacitly neglected in (5.2) and (5.4). They are proportional to the cosine of latitude and involve vertical dynamics, either in the force (II) or in the current (IV). As long as motions are predominantly horizontal, as is the case for tides, the terms II and IV are generally insignificant and can be ignored. This is known as the *Traditional Approximation*.

The Traditional Approximation goes back to Laplace's derivation of what has since been known as the *Laplace's tidal equations* (LTE). They are linear and hydrostatic like (5.14), (5.15), and (5.20), but apply to a spherical geometry in which f varies with latitude instead of being taken constant. A complete inventory of the wave solutions of the LTE, for an ocean covering the entire sphere, was made by Longuet-Higgins (1968a). For a review of the approximations underlying LTE and their validity, including the Traditional Approximation, see Hendershott (1981).

The LTE have also been used to numerically obtain free (i.e., unforced) solutions at given frequencies for particular basins, as eigenfunctions of the system. These normal modes help interpret the difference in importance of tidal constituents in the ocean basins (Figures 5.1 and 5.2), from examining the resonance characteristics of the normal modes. Notably this explains the predominance of semidiurnal tides in the Atlantic Ocean and the relative importance of diurnal tides in the Pacific Ocean (Platzman, 1984; Sanchez, 2008). An extensive comparison between different global tidal models and their assessment can be found in Stammer et al. (2014).

The problem of tides in a semi-enclosed basin was originally studied by Taylor (1922). In Section 5.6, we used the same setting and assumptions. Later studies have included other aspects as well, in particular bottom friction and variations in water depth, which make the results more applicable to real basins. Several cases are examined in Roos and Schuttelaars (2011), which also contains an extensive literature overview.

The double Kelvin wave, briefly discussed in Section 5.7.2, was first derived and examined by Longuet-Higgins (1968b). For several special cases of shelves of varying depth, solutions of continental shelf waves have been obtained, usually as nondivergent waves, i.e., on the assumption that the first term (5.20) can be neglected (LeBlond and Mysak, 1978). Observations on continental shelf waves, in the form of diurnal tides at high latitudes, are described in Lam (1999).

6

Tides in Coastal Seas and Basins

6.1 Introduction

As we move from the ocean to the coastal zone, more elements come into play in the dynamics of tides. Most of them are a direct consequence of the shallowness of the system: as the currents become stronger, bottom friction gains importance and so do the effects of nonlinearity. Moreover, the bottom boundary layer now occupies a relatively large fraction of the water column. Another element is riverine outflow of freshwater and resulting horizontal and vertical density gradients. In this chapter we first deal with topics that can be treated in a depth-averaged sense (shallow-water tidal constituents, tidal-residual currents, co-oscillation, resonance, frictional and radiation damping) and then with those in which the vertical structure is an essential part of the problem (Ekman layers, tidal straining).

6.2 Equations of Motion

We revisit the horizontal momentum equations (5.2) and (5.3), but now include a friction term

$$\frac{\partial u}{\partial t} + u\frac{\partial u}{\partial x} + v\frac{\partial u}{\partial y} + w\frac{\partial u}{\partial z} - fv = -\frac{1}{\rho}\frac{\partial p}{\partial x} + \frac{1}{\rho}\frac{\partial F_x}{\partial z} \tag{6.1}$$

$$\frac{\partial v}{\partial t} + u\frac{\partial v}{\partial x} + v\frac{\partial v}{\partial y} + w\frac{\partial v}{\partial z} + fu = -\frac{1}{\rho}\frac{\partial p}{\partial y} + \frac{1}{\rho}\frac{\partial F_y}{\partial z}, \tag{6.2}$$

where $\partial F_x/\partial z$ and $\partial F_y/\partial z$ represent stress forces.

We still assume density ρ to be constant (except in Section 6.9, where we discuss the role of density gradients in a descriptive way). Mass conservation is thus expressed by (5.7):

$$\frac{\partial u}{\partial x} + \frac{\partial v}{\partial y} + \frac{\partial w}{\partial z} = 0. \tag{6.3}$$

As a boundary condition at the free surface, we have again (5.18):

$$w = \frac{\partial \zeta}{\partial t} + u\frac{\partial \zeta}{\partial x} + v\frac{\partial \zeta}{\partial y} \qquad \text{at } z = \zeta. \tag{6.4}$$

At the bottom $z = -H(x, y)$, which we no longer assume to be flat, we replace the boundary condition (5.16) with

$$w = -u\frac{\partial H}{\partial x} - v\frac{\partial H}{\partial y} \qquad \text{at } z = -H. \tag{6.5}$$

In addition, because of the presence of the stress force, we need to prescribe the tangential stress at the surface and bottom as well. At the surface we assume that no forcing or friction is exerted on the fluid, so

$$F_x|_{z=\zeta} = 0, \quad F_y|_{z=\zeta} = 0. \tag{6.6}$$

At the bottom, we express the boundary condition as

$$F_x|_{z=-H} = \tau_x, \quad F_y|_{z=-H} = \tau_y, \tag{6.7}$$

where $\tau_{x,y}$ are to be specified in terms of the current velocities.

In the vertical, we assume the hydrostatic balance (5.12),

$$\frac{\partial p}{\partial z} = -\rho g, \tag{6.8}$$

hence

$$p = \rho g(\zeta - z) + p_a, \tag{6.9}$$

with atmospheric pressure p_a, assumed constant.

With (6.9), the momentum equations (6.1) and (6.2) become

$$\frac{\partial u}{\partial t} + u\frac{\partial u}{\partial x} + v\frac{\partial u}{\partial y} + w\frac{\partial u}{\partial z} - fv = -g\frac{\partial \zeta}{\partial x} + \frac{1}{\rho}\frac{\partial F_x}{\partial z} \tag{6.10}$$

$$\frac{\partial v}{\partial t} + u\frac{\partial v}{\partial x} + v\frac{\partial v}{\partial y} + w\frac{\partial v}{\partial z} + fu = -g\frac{\partial \zeta}{\partial y} + \frac{1}{\rho}\frac{\partial F_y}{\partial z}. \tag{6.11}$$

Like in Section 5.2.2, we vertically integrate the continuity equation (6.3), but with three modifications. First, as we have introduced frictional effects, we can no longer assume the currents u and v to be uniform in the vertical; second, we now allow H to vary in the horizontal; and third, we retain the nonlinear terms. We define the depth-averaged currents $\bar{u}$ and $\bar{v}$ as

$$\bar{u} = \frac{1}{H+\zeta}\int_{-H}^{\zeta} u\, dz, \qquad \bar{v} = \frac{1}{H+\zeta}\int_{-H}^{\zeta} v\, dz. \tag{6.12}$$

As demonstrated in Appendix B, vertical integration of (6.3) from $z = -H$ to $z = \zeta$ yields the following equation

$$\frac{\partial \zeta}{\partial t} + \frac{\partial}{\partial x}[(H + \zeta)\bar{u}] + \frac{\partial}{\partial y}[(H + \zeta)\bar{v}] = 0. \tag{6.13}$$

Vertical integration of the momentum equations (6.10) and (6.11) is less straight-forward and involves some parameterizations, as detailed in Appendix B; hereafter we use the following form:

$$\frac{\partial \bar{u}}{\partial t} + \bar{u}\frac{\partial \bar{u}}{\partial x} + \bar{v}\frac{\partial \bar{u}}{\partial y} - f\bar{v} = -g\frac{\partial \zeta}{\partial x} - \frac{1}{\rho}\frac{\tau_x}{H + \zeta} \tag{6.14}$$

$$\frac{\partial \bar{v}}{\partial t} + \bar{u}\frac{\partial \bar{v}}{\partial x} + \bar{v}\frac{\partial \bar{v}}{\partial y} + f\bar{u} = -g\frac{\partial \zeta}{\partial y} - \frac{1}{\rho}\frac{\tau_y}{H + \zeta}. \tag{6.15}$$

The bottom stress terms can be represented by

$$\tau_x = \rho C_D (\bar{u}^2 + \bar{v}^2)^{1/2}\bar{u}, \qquad \tau_y = \rho C_D (\bar{u}^2 + \bar{v}^2)^{1/2}\bar{v}, \tag{6.16}$$

where C_D is the drag coefficient (see Box 6.1).

Box 6.1 **Logarithmic Velocity Profile**

A simple model for the description of the vertical structure of a (tidal) current is the logarithmic profile

$$u = \frac{u_*}{\kappa} \log\left(\frac{z + H + z_0}{z_0}\right), \tag{6.17}$$

where κ is the Von Kármán constant (approximately 0.4), u_* the friction velocity, and z_0 the roughness length. In this context, the "bottom" (at $z = -H$) effectively refers to the top of the viscous sublayer. For tidal currents, the time dependence appears on the right-hand side of (6.17) primarily via u_*; we shall assume z_0 to be constant. Typically, u_* is of the order of a few centimeters per second, and z_0 of the order of millimeters to a few centimeters. The friction velocity, a fictitious quantity, is defined by expressing the bottom stress τ_x as

$$\tau_x = \rho |u_*| u_*.$$

Integrating (6.17) from bottom to surface gives

$$(H + \zeta)\bar{u} = \frac{u_*}{\kappa}\left[(\zeta + H + z_0) \log\left(\frac{\zeta + H + z_0}{z_0}\right) - (\zeta + H)\right]. \tag{6.18}$$

For $z_0 \ll H$ and $|\zeta| \ll H$, we can approximate (6.18) by

$$\bar{u} = \frac{u_*}{\kappa}[\log(H/z_0) - 1]. \tag{6.19}$$

Hence we can express the bottom stress in terms of the depth-averaged current as

$$\tau_x = \rho \, \frac{\kappa^2}{[\log(H/z_0) - 1]^2} \, |\bar{u}|\bar{u}, \tag{6.20}$$

which features the drag coefficient

$$C_D = \frac{\kappa^2}{[\log(H/z_0) - 1]^2}.$$

Alternatively, the Chézy coefficient C is sometimes used, with $C^2 = g/C_D$.

6.3 Shallow-Water Constituents

In Chapter 4 we discussed the principle of how compound frequencies arise from the sum and difference of the basic frequencies. This gave rise to the constituents listed in Table 4.2. We now examine how the hydrodynamics adds constituents of its own, following the same principle.

For this purpose, it suffices to consider just one of the horizontal directions, and we ignore Coriolis effects, as the focus will be on the nonlinear terms. Hence from (6.14) and (6.16):

$$\frac{\partial \bar{u}}{\partial t} + \bar{u} \, \frac{\partial \bar{u}}{\partial x} = -g \, \frac{\partial \zeta}{\partial x} - C_D \, \frac{|\bar{u}|\bar{u}}{H + \zeta}. \tag{6.21}$$

We assume the presence of two basic tidal signals (with index 1 and 2) of the form

$$\bar{u}_{1,2} = G_{1,2}(x) \exp(i\omega_{1,2}t) + \text{complex conjugate},$$

where $G_{1,2}$ is some complex function (whose form does not concern us here). The complex conjugate is added to make the signal real.

The advective term $\bar{u} \, \partial \bar{u}/\partial x$ in (6.21) thus produces the sum and difference frequencies $\omega_1 \pm \omega_2$. A special case occurs when the basic frequencies are equal: $\omega_1 = \omega_2$. For example, with M_2, which is generally the largest constituent, this leads to a time-independent term (which will be further explored in Section 6.4) and the double, quarterdiurnal frequency M_4: $\omega_{M_4} = 2 \times \omega_{M_2}$. This is a new frequency; it does not stem directly from the celestial motions, like the frequencies listed in Table 4.2, but only indirectly, as it is produced by the hydrodynamics out of the basic astronomical constituent M_2. We can proceed by combining M_2 (as ω_1) with the new M_4 (as ω_2); their sum produces $\omega_{M_6} = 3 \times \omega_{M_2}$. This is referred to as an odd constituent, as opposed to the even constituent M_4 and the next one in line: M_8. This procedure can be repeated without end, but in practice, constituents higher than M_{10} can be ignored. These shallow-water constituents, being multiples of a basic

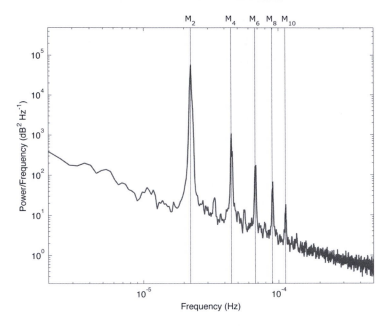

Figure 6.1 Power spectral density of the near-bottom alongshore current velocity at a location in the North Sea coastal zone (off Egmond, local water depth 11 m), showing overtides of M_2. The spectrum is based on measurements over a 16-month period, using an Acoustic Doppler Velocimeter (ADV) mounted on a bottom frame. After Van der Hout et al. (2017).

frequency, are called *overtides*. An example of overtides in an observed current record is shown in Figure 6.1.

We now take distinct astronomical constituents ω_1 and ω_2 as a starting point. Their sum and difference frequencies are called *compound tides*. For example, the sum of M_2 and S_2 frequencies produces a quarterdiurnal frequency (MS_4), while their difference leads us via (3.9) to a long-period tide whose period is half a synodic month:

$$\omega_{S_2} - \omega_{M_2} = 2\tilde{D}_{sol} - 2\tilde{D}_{lun} = 2\tilde{M}_{syn} \qquad (\text{MS}).$$

This actually reproduces the frequency of an astronomical constituent, the long-period variational MS_f (see Table 4.2).[1] In the same way, M_2 combined with N_2 creates a quarterdiurnal MN_4 and a long-period constituent MN, which equals $\tilde{M}_{ano}$ (see Table 4.2) and hence is indistinguishable from the lunar monthly elliptic M_m.

---

[1] This constituent should not be confused with the lunar fortnightly *declinational* M_f, which has a period of half a tropical month. MS/MS_f has the same period as the spring-neap cycle but truly is a long-period oscillation, whereas the spring-neap cycle is a long-period *modulation* of a semidiurnal oscillation.

We can also combine M_2 with diurnal constituents like K_1. Their sum produces a terdiurnal MK_3 (i.e., a constituent with three oscillations per day), while their difference frequency can be expressed as

$$\omega_{M_2} - \omega_{K_1} = 2\tilde{D}_{lun} - \tilde{D}^*_{sid} = \tilde{D}_{lun} - \tilde{M}_{tro} \qquad (MK_1),$$

where we used (3.10). This shallow-water constituent is indistinguishable from the astronomical constituent O_1.

The nonlinear term $\partial(\zeta\bar{u})/\partial x$ in the continuity equation (6.13) similarly produces new compound frequencies and duplicates of astronomical ones. The friction term (6.21) requires special attention because it contains an absolute current velocity as well as a cubic nonlinearity. Starting from a single semidiurnal signal $\bar{u}$, the absolute signs immediately turn $|\bar{u}|$ into a quarterdiurnal signal (plus a residual flow). Hence the combination $|\bar{u}|\bar{u}$ produces, amongst other things, a sixth-diurnal overtide (e.g., M_2 produces M_6). Starting from two different frequencies, for example M_2 and K_1, the difference and sum frequencies are the terdiurnal $2MK_3$ and fifth-diurnal $2MK_5$, respectively. In addition, the friction term contains a cubic nonlinearity, which appears as the second term in the expansion

$$\frac{|\bar{u}|\bar{u}}{H + \zeta} = \frac{|\bar{u}|\bar{u}}{H}\left(1 - \frac{\zeta}{H} + \cdots\right),$$

where we used the binomial series (A.9). Nonlinearities involving ζ are of special importance in the interaction of storm surges with tides, whereas the advective term is relatively insignificant in this respect. Storm surges lack the neat periodicity of tidal constituents, and hence they smear the spectral lines.

Finally, we mention two more compound tides that have a special significance. First, the combination

$$\omega_{M_2} + \omega_{M_2} - \omega_{S_2} = 2\tilde{D}_{lun} - 2\tilde{M}_{syn} \qquad (2MS_2),$$

where we used (3.9). This replicates the frequency of the astronomical larger variational semidiurnal constituent μ_2, listed in Table 4.2. In tide-gauge records, this constituent can therefore appear to be more prominent than expected on the basis of the ranking in the tide-generating potential. The same is true for the smaller lunar elliptic semidiurnal L_2, whose frequency is replicated by

$$\omega_{M_2} + \omega_{M_2} - \omega_{N_2} = 2\tilde{D}_{lun} + \tilde{M}_{ano} \qquad (2MN_2).$$

Thus, there will effectively be a transfer of energy from M_2 and N_2 to L_2, which explains why, in Figure 4.4, L_2 is larger than expected in relation to N_2.

We illustrate the importance of shallow-water constituents by returning once again to the tide-gauge record of Figure 1.13. The largest constituents in this record are plotted in Figure 6.2. In addition to the species of long-period, diurnal, and

Table 6.1 *List of main shallow-water constituents.*

Name	Frequency (°/hour)	Type	Origin	Astronomical equivalent
MS	1.015896	long-period	S_2-M_2	MS_f
MN	0.544375	long-period	M_2-N_2	M_m
MK_1	13.943036	diurnal	M_2-K_1	O_1
MO_1	15.041069	diurnal	M_2-O_1	K_1
$2MN_2$	29.528479	semidiurnal	$M_2+M_2-N_2$	L_2
$2MS_2$	27.968208	semidiurnal	$M_2+M_2-S_2$	μ_2
MK_3	44.025173	terdiurnal	M_2+K_1	$-$
MO_3	42.927140	terdiurnal	M_2+O_1	$-$
$2MK_3$	42.927140	terdiurnal	$M_2+M_2-K_1$	$-$
M_4	57.968208	quarterdiurnal	M_2+M_2	$-$
MS_4	58.984104	quarterdiurnal	M_2+S_2	$-$
MN_4	57.423834	quarterdiurnal	M_2+N_2	$-$
$2MK_5$	73.009277	fifth-diurnal	$M_2+M_2+K_1$	$-$
M_6	86.952313	sixth-diurnal	$M_2+M_2+M_2$	$-$
$2MS_6$	87.968208	sixth-diurnal	$M_2+M_2+S_2$	$-$

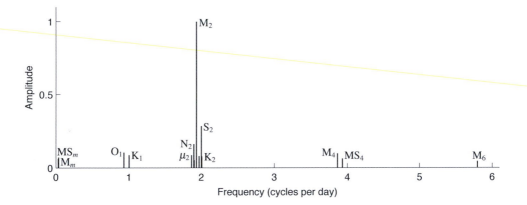

Figure 6.2 Normalized amplitudes of the constituents obtained by harmonic analysis from the tide-gauge record shown in Figure 1.13.

semidiurnal tides, we have quarterdiurnal and sixth-diurnal ones. We see a relatively prominent μ_2 and L_2 (which is the line between M_2 and S_2). MS_m is the long-period evectional constituent listed in Table 4.2.

The astronomical constituents are primarily generated in the oceans and from there spread onto the continental shelves (see, e.g., Figure 5.12) and into coastal waters. For the shallow-water constituents, it is the other way round: they originate in the shallow seas where nonlinearities are strong, and from there they spread into the oceans, creating complicated patterns of interference of signals from different

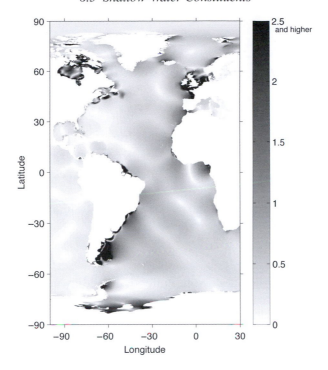

Figure 6.3 Map of the overtide M_4 in the Atlantic Ocean, showing the amplitude in centimeters. Figure generated using Aviso+ products, courtesy of LEGOS/Noveltis/CNES/CLS.

sources (Figure 6.3). This illustrates the important point that the propagation of tidal energy from the oceans onto continental shelves is not a one-way street; to a certain degree, the signal in the ocean is affected by the dynamics on the shelf.

To conclude this section, we consider a superposition of a basic semidiurnal constituent (here: M_2) and its first overtide (M_4). We write the signal as

$$\bar{u} = A \sin(\omega_{M_2} t) + B \sin(\omega_{M_4} t + \varphi). \tag{6.22}$$

We take $A = 1$ and $B = 0.25$ (in m/s) and examine two choices for the phasing φ, shown in Figure 6.4. They create distinct kinds of asymmetry. In Figure 6.4a, the ebb and flood currents are different in strength as a result of the superposition of M_2 and M_4: the current at maximum ebb is stronger than at maximum flood. In coastal areas, such an asymmetry generally affects the transport of sediment. Erosion of fine sediment increases with the cubic power of the current strength, so this greatly favors (in this case) erosion and transport during ebb. The asymmetry of the tide thus translates into an asymmetry in sediment transport. In Figure 6.4b the phasing is different, resulting in an equality of maximum flood and ebb. However, there is now another asymmetry: the time interval from maximum flood to maximum ebb is considerably longer than the other way round. This again affects the transport of fine

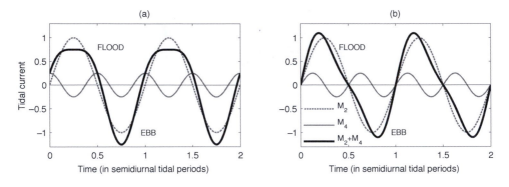

Figure 6.4 Superposition of M$_2$ and its first overtide M$_4$, for different phasing: in (a), $\varphi = \pi/2$; in (b), $\varphi = 0$.

sediment: after resuspension at maximum flood, the sediment has more time to settle to the bottom than after maximum ebb, when it soon moves back with the flood without having had the time to settle. This favors transport in the flood direction.

The phasing between the constituents generally shows a strong spatial variation in coastal areas, because of the complex way of tidal propagation and the local generation of shallow-water constituents. This often results in markedly different curves of the tidal currents (and also of surface elevation) even within the same region, as well as different characteristics of (net) sediment transport.

Exercise

6.3.1   Can the signal of the long-period constituent MS$_m$ in tide-gauge records (e.g., in Figure 6.2) be partly due to shallow-water constituents?

6.4 Tidally Induced Residual Flows

In the previous section we briefly touched upon the generation of a time-independent term as a result of nonlinear interaction. The implication is that a purely oscillatory tidal current gives rise to a net current. In other words, after averaging over a tidal cycle, one finds a residual signal. The approach in the previous section was provisional: we explored just what comes out from multiplications of constituents in the nonlinear terms without considering the full equations or a specific setting. The latter is necessary in order to identify the conditions under which such a transfer of energy actually becomes possible.

Here, we examine those conditions for a simple setting and also explore some implications of tide-induced residual currents for the transport of water parcels.

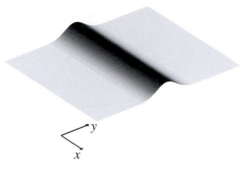

Figure 6.5 Setting of the problem in Section 6.4.1: a sand bank positioned along the x-axis (extending to infinity in both directions) and varying along the y-axis.

6.4.1 Residual Flow along a Bank

Observations and numerical model results point to a central role of the bathymetry in the generation of residual currents.[2] We therefore choose a setting that features a sand bank (Figure 6.5), described by $z = -H(y)$, for a certain function H, the local water depth. As before, the free surface lies at $z = \zeta$.

The bank stretches uniformly along the x axis. To simplify the analysis, we assume that the horizontal depth-averaged currents $\bar{u}$ and $\bar{v}$, too, are uniform in x. Far away from the bank (the far field, denoted by subscript ∞), we impose a spatially uniform tidal flow

$$(\bar{u}_\infty, \bar{v}_\infty) = (s, 1)\, V \sin \omega t, \tag{6.23}$$

with constants s and V. This describes a rectilinear tidal current under a certain angle with the bank, which is specified by s. For $s = 0$, the tidal flow is normal to the bank, whereas for large $|s| \gg 1$, it becomes aligned to the bank.

As the currents are independent of x, the depth-averaged horizontal momentum momentum equations (6.14) and (6.15) become

$$\frac{\partial \bar{u}}{\partial t} + \bar{v}\frac{\partial \bar{u}}{\partial y} - f\bar{v} = -g\frac{\partial \zeta}{\partial x} - \frac{r\bar{u}}{H} \tag{6.24}$$

$$\frac{\partial \bar{v}}{\partial t} + \bar{v}\frac{\partial \bar{v}}{\partial y} + f\bar{u} = -g\frac{\partial \zeta}{\partial y} - \frac{r\bar{v}}{H}. \tag{6.25}$$

Here, we made another approximation to simplify the analysis: we have replaced the nonlinear friction term by a linear one, which features a coefficient r that is assumed to be constant. Notice that the free surface ζ may depend (linearly) on x, so that its gradient can drive the flow.

---

[2] In particular, the study by Huthnance (1973), whose analysis we follow here, was inspired by observations over the Norfolk sandbanks.

In the present setting, the continuity equation (6.13) becomes

$$\frac{\partial \zeta}{\partial t} + \frac{\partial}{\partial y}[(H + \zeta)\bar{v}] = 0. \tag{6.26}$$

Throughout this analysis, we assume that $|\zeta| \ll H$. For sufficiently narrow banks, the second term in (6.26) will be dominant and the flow over the bank is essentially governed by

$$\frac{\partial}{\partial y}(H\bar{v}) = 0. \tag{6.27}$$

Hence $H\bar{v}$ is constant, and with the far-field condition (6.23), this yields $\bar{v}$:

$$\bar{v}(t, y) = \frac{H_\infty \bar{v}_\infty(t)}{H(y)} = \frac{H_\infty V \sin \omega t}{H(y)}. \tag{6.28}$$

Thus, the cross-slope flow $\bar{v}$ is everywhere oscillatory and contains no residual component.

We now derive the solution for the along-slope flow $\bar{u}$. First, we take the derivative of (6.25) with respect to x; this makes all terms vanish except

$$\frac{\partial}{\partial x}\left(\frac{\partial \zeta}{\partial y}\right) = 0.$$

Hence, by interchanging the derivatives, it follows that $\partial \zeta / \partial x$ must be uniform in y. We exploit this result by evaluating the momentum equation (6.24) at two different locations of y, at an arbitrary near-bank location y and in the far field y_∞, and then subtract the results. Since $\partial \zeta / \partial x$ is the same at both locations, the term disappears after subtraction and hence we obtain

$$\frac{\partial}{\partial t}\left(\bar{u} - \bar{u}_\infty\right) + \bar{v}\frac{\partial \bar{u}}{\partial y} - f(\bar{v} - \bar{v}_\infty) = -r\left(\frac{\bar{u}}{H} - \frac{\bar{u}_\infty}{H_\infty}\right). \tag{6.29}$$

Notice that there is no term $\bar{v}_\infty \partial \bar{u}_\infty / \partial y$, since the current velocity is spatially uniform in the far field. We bring all the terms involving the unknown $\bar{u}$ to the left-hand side and put the remaining terms on the right-hand-side,

$$\frac{\partial \bar{u}}{\partial t} + \bar{v}\frac{\partial \bar{u}}{\partial y} + \frac{r\bar{u}}{H} = \frac{\partial \bar{u}_\infty}{\partial t} + f(\bar{v} - \bar{v}_\infty) + \frac{r\bar{u}_\infty}{H_\infty}. \tag{6.30}$$

With $(\bar{u}_\infty, \bar{v}_\infty)$ prescribed by (6.23) and $\bar{v}$ by (6.28), we can write (6.30) as

$$\frac{\partial \bar{u}}{\partial t} + \frac{H_\infty}{H} V \sin \omega t \frac{\partial \bar{u}}{\partial y} + \frac{r\bar{u}}{H} = s V \omega \cos \omega t + q V \sin \omega t, \tag{6.31}$$

where we introduced the short hand notation

$$q = f\left(\frac{H_\infty}{H} - 1\right) + \frac{rs}{H_\infty}.$$

We solve (6.31) by writing $\bar{u}$ in the form of a series of harmonics in ωt, starting with the basic harmonic $\bar{u}_1$:

$$\bar{u} = \overbrace{A\sin\omega t + B\cos\omega t}^{\bar{u}_1} + \bar{u}_{res} + \overbrace{C\sin 2\omega t + D\cos 2\omega t}^{\bar{u}_2} + \cdots. \tag{6.32}$$

The coefficients (A, etc.) are functions of y. The series can be continued indefinitely, but we shall here derive the residual flow obtained by including just the basic harmonic $\bar{u}_1$ and the residual $\bar{u}_{res}$ (the inclusion of the second harmonic $\bar{u}_2$ would modify the residual flow but is not essential for demonstrating its existence per se).

Considering first $\bar{u}_1$ alone and ignoring the nonlinear term, (6.31) implies

$$\omega(A\cos\omega t - B\sin\omega t) + \frac{r}{H}(A\sin\omega t + B\cos\omega t) = sV\omega\cos\omega t + qV\sin\omega t.$$

The coefficients of the cosine and sine terms each have to vanish, hence we obtain an algebraic set of equations for A and B:

$$\omega A + \frac{r}{H}B = sV\omega, \qquad \frac{r}{H}A - \omega B = qV.$$

From this set, A and B can be readily solved. In particular,

$$A = \frac{s\omega^2 + qr/H}{\omega^2 + (r/H)^2}V. \tag{6.33}$$

Notice that A depends on y via H and q.

The tidal-averaged version of (6.31) is

$$\bar{u}_{res} = -\frac{H_\infty V}{r}\left\langle \sin\omega t \frac{\partial\bar{u}}{\partial y}\right\rangle, \tag{6.34}$$

which is valid up to any order in the series (6.32); here the brackets stand for tidal averaging. Considering just $\bar{u}_1$, the only contribution to the right-hand side of (6.34) comes from the sine term, hence with (A.7)

$$\bar{u}_{res} = -\frac{H_\infty V}{2r}\frac{dA}{dy}. \tag{6.35}$$

Schematically, the solution (6.35) is shown in Figure 6.6. The residual current is oriented along the bank, in opposite directions at the two sides of the bank.

From the procedure that led to (6.35), it is clear that the friction term is instrumental in generating the residual flow; this is also demonstrated by the presence of

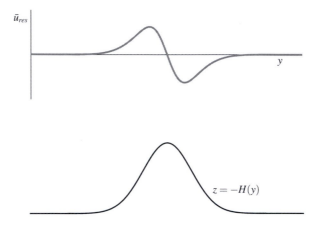

Figure 6.6 The structure of the residual current (6.35), in gray, over a bank (in black). Depending on the parameter values (including s), the sign of $\bar{u}_{res}$ can also be mirrored.

the coefficient $1/r$. However, the resulting residual flow is actually not inversely proportional to r, because the derivative dA/dy contains a factor r:

$$\frac{dA}{dy} = r \frac{[\omega^2 + (r/H)^2]\frac{d}{dy}(q/H) - r[s\omega^2 + qr/H]\frac{d}{dy}(1/H^2)}{[\omega^2 + (r/H)^2]^2} V. \qquad (6.36)$$

In the limit of vanishing friction, the Coriolis parameter f becomes (via q) the determining factor in the residual flow.

In conclusion, three elements are found to be essential for the generation of a residual current: a bottom slope, friction, and the nonlinear (i.e., advective) terms. The latter means that the effect increases quadratically with the strength of the basic tidal flow ($\bar{u}_{res} \sim V^2$). Finally, Coriolis effects also affect the residual flow. All these elements are generally present in coastal regions, which explains why tidally driven residual flows are a ubiquitous phenomenon. In practice, other mechanisms for driving residual flows often act concurrently, notably as wind and density-driven flows.

6.4.2 Residual Flows in Back-Barrier Basins

We show an example of tidal residual flows in a realistic numerical model calculation (Figure 6.7), where we have selected a period of very low wind speeds (not exceeding 2 Bft), so that the signal will be predominantly tidal, although freshwater influences may play a role close to the mainland. The residual currents generally follow the bathymetry and flow along the slopes of the channels, as one would expect from the analysis in the previous section. Near the inlet the bathymetry becomes more complex, for this is where the main channel splits up into smaller

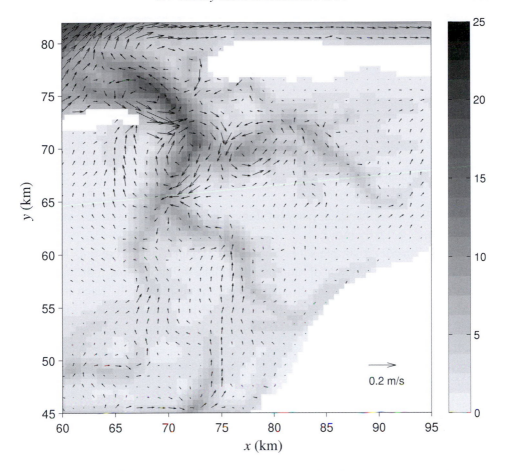

Figure 6.7 Model results on tidal residual currents in a back-barrier basin, here represented by their depth-averaged values. Land is marked in white, with the mainland at the bottom of the figure and parts of two barrier islands at the top. In between the islands, a tidal inlet is present with a deep channel, which splits up inside the basin. The bathymetry is indicated in gray scale (depth in meters). This is part of a model run of a larger domain, the inlet shown here represents the Vlie (Dutch Wadden Sea, with the domain rotated $17°$ in the counterclockwise direction with respect to the west–east axis). Based on Duran-Matute et al. (2016).

branches. At this point we observe *residual cells*: more or less closed, circular forms of circulation. Cells like this are a typical phenomenon near tidal inlets. Their significance will be further explored in an idealized setting in the following section.

6.4.3 Chaotic Stirring

In Sections 6.4.1 and 6.4.2, we dealt with the tidally driven residual current from an Eulerian perspective: a flow field as a function of the spatial coordinates. This

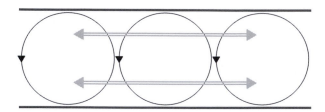

Figure 6.8 Schematic setting of tidal motion in a coastal channel-like basin (top view): a combination of an oscillatory motion (long horizontal arrows) and circular cells of residual circulation.

does not tell us how the water parcels (along with dissolved substances, suspended sediment, etc.) actually move.

We here consider a simple schematic setting depicted in Figure 6.8. The water parcels experience two kinds of motions: the tidal oscillatory flow indicated by the long horizontal arrows, and the circular flow of the residual cells. Each on its own would render the problem easy: the parcels would just move back and forth, or go around in a steady circular motion, respectively. However, as we shall see, their combined effect is nontrivial.

We first put the schematic setting of Figure 6.8 in a mathematical form:

$$u = \quad U_r \sin kx \cos ky + U_0 \sin \omega t \qquad (6.37)$$

$$v = -U_r \cos kx \sin ky, \qquad (6.38)$$

where x is the along-channel direction, with current component u, and y the cross-channel direction, with current component v; U_r is the amplitude of the residual flow, U_0 the amplitude of the oscillatory tidal flow; π/k represents the cell size, and ω the tidal frequency.

We rephrase (6.37) and (6.38) in terms of nondimensional parameters by introducing new variables

$$(x, y) = \frac{\pi}{k}(x', y'), \quad t = \frac{2\pi}{\omega}t', \quad (u, v) = \frac{\pi/k}{2\pi/\omega}(u', v') = \frac{\omega}{2k}(u', v').$$

Hence

$$u' = \quad \mu\eta \sin \pi x' \cos \pi y' + \mu \sin 2\pi t' \qquad (6.39)$$

$$v' = -\mu\eta \cos \pi x' \sin \pi y'. \qquad (6.40)$$

Hereafter, we drop the primes, for convenience. Equations (6.39) and (6.40) feature two nondimensional parameters,

$$\mu = \frac{2U_0 k}{\omega} = \frac{\text{tidal excursion}}{\text{cell size}}, \qquad \eta = \frac{U_r}{U_0} = \frac{\text{residual velocity}}{\text{tidal velocity}}. \qquad (6.41)$$

(In μ, we interpret $1/k$ loosely as the "cell size" – strictly speaking, it is given by π/k.) Parameter μ is a measure of how many cells are crossed in the oscillatory tidal motion, whereas η is a measure of the relative strength of the residual circulation. For simplicity, we regard the parameters μ and η as independent, but in a real physical setting they are connected because the residual cells are generated by the tidal oscillatory flow (in combination with the bathymetry).

In a Lagrangian view, x and y are positions of the water parcels, whose velocity is given by

$$\frac{dx}{dt} = \mu\eta \sin\pi x \cos\pi y + \mu \sin 2\pi t \qquad (6.42)$$

$$\frac{dy}{dt} = -\mu\eta \cos\pi x \sin\pi y. \qquad (6.43)$$

For an arbitrary initial position (x_0, y_0) at $t = 0$, the system can be (numerically) integrated, which yields the trajectory of the water parcel as a function of time.

Before we show an example from such a calculation, we examine the character of the dynamical system (6.42) and (6.43) more closely. First of all, we notice that the unknowns x and y appear in the arguments of the cosine and sine on the right-hand sides; hence the equations are nonlinear. Second, (6.42) features an explicit time dependence due to the presence of the oscillatory term on the right-hand side; effectively, time enters as a third dimension, in addition to x and y. A nonlinear dynamical system of three dimensions (or higher), like (6.42) and (6.43), offers the prospect of chaotic behavior. The primary characteristic of chaos is a sensitivity on the initial conditions. In other words, we may expect that for certain parameters and certain areas, initially nearby water parcels disperse rapidly.

Two examples are shown in Figure 6.9. Initially nearby particles stay together in the case of Figure 6.9a; in this case there is no sensitivity on the initial conditions. However, the particles appear to disperse quickly when released near the boundaries, as in Figure 6.9b; after six tidal periods, they have already spread over three different cells. Despite the utter simplicity of the underlying velocity field (6.37) and (6.38), the trajectories of the particles are complex and apparently chaotic.

It should be noticed that the system considered here is non-diffusive and reversible: the spreading of particles here is a form of *stirring*. The equations remain valid if one reverses time, in which case the particles in Figure 6.9b would come together again. However, if gradients exist, for example in salinity due to a local discharge of freshwater, the stirring will act to stretch the area of freshwater and thus enhance gradients at its perimeter. In any real system, this enhances diffusivity and hence facilitates mixing.

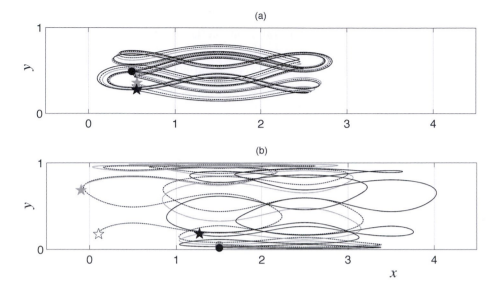

Figure 6.9 Trajectories of three particles, whose initial (final) positions are indicated by circles (asterisks). The duration is six tidal periods and parameter values are $\eta = 0.25$ and $\mu = 7$. Two cases are shown: (a) a release of particles near the center of a residual cell; (b) a release near the boundary of the channel. The dimensions of the basin are scaled, the residual cells having a size of 1×1 in this graph.

6.5 Co-Oscillation and Resonance

Lagoons, back-barrier basins, and estuaries owe their tidal dynamics to the signal of the tidal wave that passes along in the adjacent sea; they co-oscillate with that tidal wave. The response in the basin may be relatively strong, depending on its resonance characteristics. We examine this problem for the simple configuration shown in Figure 6.10.[3]

We return to the propagation of a single Kelvin wave (Section 5.4), which we now confine to a narrow channel. The notion of narrowness is here meant in the following way. We saw in Section 5.4 that the Kelvin decays off the wall at a spatial scale given by the Rossby radius of deformation, c_0/f (with $c_0 = (gH)^{1/2}$, water depth H). If we choose the channel to be narrow in the sense that $L \ll c_0/f$, then the wave stays essentially uniform across the channel. This effectively removes all traces of Coriolis effects, for in the along-channel direction, the propagation was already governed by a dispersion relation without f, as indicated by (5.45). Thus dispensing with Coriolis effects, the wave solution in the channel becomes

$$\zeta_c = \zeta_0 \exp i (ly - \omega t), \tag{6.44}$$

---

[3] In this section we largely follow the exposition given by Zimmerman (1993) in his lecture notes.

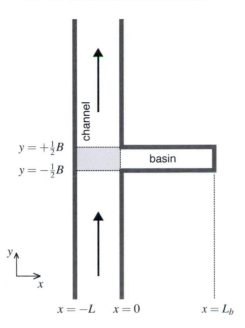

Figure 6.10 Schematic setting of a basin (width B, length L_b) connected to a channel of width L (top view). The principal tidal wave propagates through the channel in the direction indicated by the arrows. By the assumptions of narrow widths of channel and basin, as specified in the analysis, the surface elevation of the tidal wave(s) can be regarded as uniform in the gray area in front of the basin.

where ζ_0 is an arbitrary complex constant, without any dependence on x. The real part of the right-hand side is implied. The index c stands for "channel."

The channel has an opening at one side, connecting it to a narrow basin (Figure 6.10). Here, the "narrowness" is defined with respect to the wavelength of the basic tidal wave in the channel. We assume the width B to be much smaller than this wavelength, $B \ll 2\pi/l$. This means that, at any instant, the level of the passing wave (6.44) is fundamentally uniform along the entrance of the basin. So far as the basin is concerned, we can thus assume uniformity in y. Moreover, we assume that the water depth is constant (H, as in the channel). With $f = 0$, $\partial/\partial y = 0$, and ignoring nonlinear terms and friction, (6.13) and (6.14) reduce to

$$\frac{\partial \zeta_b}{\partial t} + H \frac{\partial u_b}{\partial x} = 0 \tag{6.45}$$

$$\frac{\partial u_b}{\partial t} = -g \frac{\partial \zeta_b}{\partial x}, \tag{6.46}$$

where the subscript b stands for "basin." Throughout this section, we consider currents to be depth-averaged, but leave out the bars over u and v, for simplicity of notation. Combined, (6.45) and (6.46) imply

$$\frac{\partial^2 u_b}{\partial t^2} - c_0^2 \frac{\partial^2 u_b}{\partial x^2} = 0. \tag{6.47}$$

The channel wave (6.44) imposes its periodicity on the co-oscillating response in the basin, hence we can express u_b as

$$u_b = u_0(x) \exp(-i\omega t), \tag{6.48}$$

for a certain complex function $u_0(x)$. Substitution in (6.47) gives

$$u_0'' + k^2 u_0 = 0, \tag{6.49}$$

where the prime stands for the derivative to x. We introduced wavenumber $k = \omega/c_0$ (which actually equals l, but to retain a clear distinction between the x and y directions, we use k and l separately). The general solution of (6.49) is

$$u_0 = C_1 \sin k(x - L_b) + C_2 \cos k(x - L_b). \tag{6.50}$$

We have added a phase shift kL_b in the argument for later convenience; this involves no loss in generality, since the complex constants $C_{1,2}$ are still unspecified. They follow from the boundary conditions for u_b, here applied to u_0. At the end of the basin ($x = L_b$), we naturally impose $u_0 = 0$, hence $C_2 = 0$. At the entrance ($x = 0$), we prescribe $u_0 = U$, for a certain complex constant U (to be determined later). With that, $C_1 = -U/\sin kL_b$ and so u_b takes the form

$$u_b = -\frac{U}{\sin kL_b} \sin k(x - L_b) \exp(-i\omega t). \tag{6.51}$$

We obtain the corresponding free surface elevation in the basin from (6.45),

$$\zeta_b = i \frac{HU}{c_0 \sin kL_b} \cos k(x - L_b) \exp(-i\omega t). \tag{6.52}$$

Finally, we impose the condition that the elevation at the entrance of the basin equals the neighboring ζ_c in the channel,

$$\zeta_c|_{y=0} = \zeta_b|_{x=0}. \tag{6.53}$$

Evaluating this equality, with ζ_c given by (6.44), we obtain an expression for U:

$$U = \frac{c_0 \zeta_0 / H}{i \cot kL_b}. \tag{6.54}$$

With this, the problem has been solved. In particular, we can now write ζ_b in (6.52) as

$$\zeta_b = \frac{\zeta_0}{\cos k L_b} \cos k(x - L_b) \exp(-i\omega t). \tag{6.55}$$

With a finite signal at the entrance of the basin, we can nevertheless have an infinite response at its end, namely when $\cos k L_b = 0$, which is the case for

$$k L_b = \frac{\pi}{2} + n\pi, \quad \text{for } n = 0, 1, 2 \cdots.$$

In terms of the wavelength $\lambda = 2\pi/k$, this becomes

$$L_b = \left(\frac{1}{4} + \frac{n}{2}\right)\lambda, \quad \text{for } n = 0, 1, 2 \cdots, \tag{6.56}$$

stating that resonance occurs when the basin length is one quarter (or three quarters, etc.) of the wavelength. An infinite response is plainly unphysical, so we need to find ways to remedy this behavior by including effects that we have so far neglected but that are in fact pertinent to the problem at hand. One obvious candidate is the inclusion of friction. After all, close to resonance, not only the elevation ζ_b tends to infinity, but so do U and hence u_b. Frictional terms like (6.16) then inevitably become strong and dampen the response. We discuss this mechanism of *frictional damping* in Section 6.5.1.

There is yet another mechanism to quench the response in the basin. So far, we have treated the occurrence of resonance as something that concerns the basin only. In particular, the sea level at the entrance was specified by the tidal wave in the channel, irrespective of what happens in the basin. More realistically, we expect that in a state close to resonance, the response will not be confined to the basin but will literally radiate out into the channel. In other words, the open boundary at $x = 0$ becomes a two-way street: one that allows a co-oscillation with the channel, but also a release of energy from the basin into the channel, depending on the state of resonance. The level at the entrance is then no longer known a priori but becomes part of the problem. This mechanism of *radiation damping* is explored in Section 6.5.2.

6.5.1 Frictional Damping

We modify (6.46) by adding a linear frictional term on its right-hand side, like in (6.24),

$$\frac{\partial u_b}{\partial t} = -g \frac{\partial \zeta_b}{\partial x} - \frac{r u_b}{H}, \tag{6.57}$$

with a certain constant r, which will be specified later. Combining (6.45) and (6.57) gives

$$\frac{\partial^2 u_b}{\partial t^2} - c_0^2 \frac{\partial^2 u_b}{\partial x^2} = -\frac{r}{H}\frac{\partial u_b}{\partial t}. \tag{6.58}$$

We still assume that the tidal periodicity is imposed by the channel. Hence, with (6.48),

$$u_0'' + \kappa^2 u_0 = 0. \tag{6.59}$$

This equation is formally the same as (6.49), but with k replaced by κ, which is defined as

$$\kappa = k\left(1 + \frac{ir}{c_0 k H}\right)^{1/2}.$$

The procedure is the same as before, except that we are dealing with the complex κ instead of k. As a result, ζ_b is now given by

$$\zeta_b = \frac{\zeta_0}{\cos \kappa L_b}\cos\kappa(x - L_b)\exp(-i\omega t). \tag{6.60}$$

The friction coefficient r can be specified by assuming equality with $C_D U_s$, where U_s is the typical velocity scale, here taken to be 1 m/s. A typical value of C_D is 2.5×10^{-3}.

We show examples for various strengths of friction in Figure 6.11. The resonant peaks become smaller for stronger friction, as expected. Moreover, they shift to slightly smaller wavenumbers as friction increases. It can also be seen that the peak

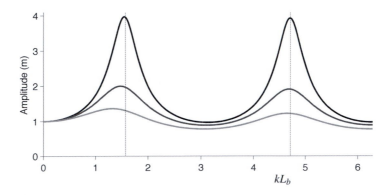

Figure 6.11 An illustration of frictional damping. The amplitude of ζ_b at the end of the basin, obtained by taking the absolute value of (6.60) at $x = L_b$, plotted against the scaled wavenumber kL_b. Parameter values are $\zeta_0 = 1$ m, $H = 10$ m and $L_b = 20$ km. Friction increases from $r = 2.5 \times 10^{-3}$ m/s (black line), to $r = 5 \times 10^{-3}$ m/s (gray line) and $r = 7.5 \times 10^{-3}$ m/s (light gray line). The vertical lines indicate the one-quarter wavelength resonance (left) and three-quarter wavelength resonance (right), defined in (6.56).

at one-quarter wavelength undergoes a larger shift than the one at three-quarter wavelength. On the other hand, the latter is more reduced by friction than the former.

6.5.2 Radiation Damping

We now ignore friction but allow waves to radiate out from the basin into the channel. We denote these waves by ζ_r and v_r. They are induced by an exchange flow between channel and basin, whose volume flux F is given by the cross-sectional area of the entrance (i.e., basin width B times water depth H) multiplied by the current velocity at the entrance:

$$F = BHU \exp(-i\omega t).$$

We note that U is no longer given by (6.54) but will take a different form in the present setting. Because of the symmetry of the problem, the flow will be evenly distributed into (or withdrawn from) the channel through the boundaries at $y = \pm B/2$, each having a cross-sectional area LH. Hence

$$v_r = \pm \frac{BU}{2L} \exp(-i\omega t) \quad \text{at } y = \mp B/2. \tag{6.61}$$

A check of the signs is in order: for positive U (i.e., a flow into the basin), v_r is positive at $y = -B/2$ and negative at $y = +B/2$, both correctly representing a flow into the area in front of the basin.

We obtain an equation for v_r by returning to (6.13) and (6.15); ignoring friction, nonlinear terms and Coriolis effects, and setting $\partial/\partial x = 0$, it follows that v_r satisfies

$$\frac{\partial^2 v_r}{\partial t^2} - c_0^2 \frac{\partial^2 v_r}{\partial y^2} = 0. \tag{6.62}$$

Its general solution has waves propagating in both directions, but in the present setting we naturally impose that the radiating waves travel *away* from the source, i.e., in the negative y-direction for $y < -B/2$ and in the positive y-direction for $y > B/2$. Thus

$$v_r = \frac{BU}{2L} \times \begin{cases} \exp i(-l[y+B/2] - \omega t) & \text{for } y < -B/2 \\ -\exp i(+l[y-B/2] - \omega t) & \text{for } y > +B/2, \end{cases} \tag{6.63}$$

where appropriate coefficients have been chosen to accommodate the boundary conditions (6.61). The corresponding free surface elevation follows from (6.13), under the same assumptions as those leading to (6.62),

$$\zeta_r = -\frac{BHU}{2Lc_0} \times \begin{cases} \exp i(-l[y+B/2] - \omega t) & \text{for } y < -B/2 \\ \exp i(+l[y-B/2] - \omega t) & \text{for } y > +B/2. \end{cases} \tag{6.64}$$

At the boundaries $y = \pm B/2$, ζ_r is identical, and this is taken as the uniform value within the gray area in front of the basin (Figure 6.10).

Finally, we impose the condition that the elevation at both sides of the entrance of the basin be equal. Instead of (6.53), we now have

$$\zeta_c|_{y=0} + \zeta_r|_{y=0} = \zeta_b|_{x=0}. \tag{6.65}$$

With ζ_c from (6.44), ζ_b from (6.52), and ζ_r from (6.64), this becomes

$$\zeta_0 - \frac{BHU}{2Lc_0} = i \frac{HU}{c_0 \tan kL_b},$$

from which U is solved:

$$U = \frac{c_0\zeta_0/H}{B/(2L) + i \cot kL_b}. \tag{6.66}$$

This new expression for U differs crucially from (6.54) in that there is no longer an infinite response for $\cos kL_b = 0$. It has been quenched by *radiation damping*, i.e., by allowing waves to radiate out of the basin. Substitution of U from (6.66) in (6.52) gives the tidal motion of the free surface in the basin,

$$\zeta_b = i \frac{\zeta_0 \cos k(x - L_b)}{[B/(2L)] \sin kL_b + i \cos kL_b} \exp(-i\omega t). \tag{6.67}$$

In Figure 6.12 we illustrate two cases for different basin widths. Resonance peaks still occur, but now have a finite amplitude. Radiation damping differs from frictional damping (discussed in the previous section) in two fundamental respects. The resonance peaks are now not shifted, but only reduced. Consequently, they stay precisely at one-quarter wavelength, three-quarter wavelength, etc. Moreover,

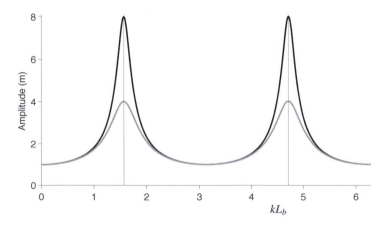

Figure 6.12 The amplitude of ζ_b at the end of the basin, obtained by taking the absolute value of (6.67) at $x = L_b$, plotted as a function of kL_b. Parameter values are $\zeta_0 = 1$ m and $L = 20$ km. Two values of the basin width B are shown: 5 km (black) and 10 km (gray).

the two peaks are damped in the same way, in contrast to the case of frictional damping, where they became different.

Exercise

6.5.1 Examine the combined effects of frictional and radiation damping. What is their relative importance?

6.6 Helmholtz Oscillation

A type of resonance with different characteristics than in the previous section exists in the setting depicted in Figure 6.13. Again, we consider the principal tidal wave to propagate in the channel, but the channel is now connected to a lagoon, via a narrow inlet. We assume that the surface in the lagoon (ζ_l) moves *uniformly* up and down in response to the oscillating level at its entrance. This is called a *Helmholtz oscillation*. Let the area of the lagoon be A; we can then specify the volume flux F through the inlet as

$$F = A \frac{d\zeta_l}{dt}. \tag{6.68}$$

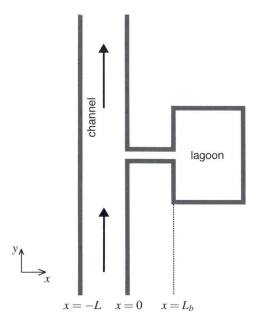

Figure 6.13 Schematic setting of a lagoon (area A) connected to a channel of width L, via a narrow inlet (top view). The principal tidal wave propagates through the channel in the direction indicated by the arrows.

In the inlet, we have a pressure gradient specified by the difference in water level between the lagoon (ζ_l) and the channel (ζ_c), which drives the flux F. We assume the slope in water level to be linear throughout the inlet. Dividing the flux by the cross-sectional area of the inlet (water depth H times width B), we obtain the current in the inlet: $u_b = F/(HB)$. The momentum equation (6.46) thus becomes

$$\frac{1}{HB}\frac{dF}{dt} = -g\frac{\zeta_l - \zeta_c}{L_b}. \tag{6.69}$$

Combining (6.68) and (6.69) gives

$$\frac{AL_b}{gHB}\frac{d^2\zeta_l}{dt^2} + \zeta_l = \zeta_c. \tag{6.70}$$

The oscillation at the entrance of the inlet, $\zeta_c = \zeta_0 \exp(-i\omega t)$, imposes its periodicity on the dynamics of the lagoon, hence we seek a solution of the form

$$\zeta_l = \hat{\zeta}_l \exp(-i\omega t).$$

Substitution in (6.70) gives the amplitude of the oscillation in the lagoon,

$$\hat{\zeta}_l = \frac{\zeta_0}{1 - \omega^2 AL_b/(gHB)}. \tag{6.71}$$

Resonance now occurs when

$$\omega = \left(\frac{gHB}{AL_b}\right)^{1/2}. \tag{6.72}$$

For realistic values of the various parameters, this may well be within the range of tidal frequencies. For example, the basin of the Strait of Georgia (east of Vancouver Island), with an estimated area of 9,000 km^2, is connected to the adjacent bay via narrow channels with a length of about 25 km and a combined cross-sectional area (i.e., HB) of 0.6 km^2. Using these values in (6.72), one finds a frequency of slightly less than 11 h, which is close to semidiurnal.

Exercise

6.6.1   Examine the role of frictional damping in the Helmholtz oscillation by adding a linear friction term.

6.7 Tidal Currents: Decomposition in Phasors

Horizontal tidal currents belonging to a specific constituent can be depicted as ellipses in the uv-plane, as for example in Figures 5.11 and 5.15. The ellipses come in different shapes, from rectilinear to circular. Moreover, ellipses can be tilted with respect to the uv-axes, which means that maximum currents are at an angle with the uv-axes. We insert this technical section to describe a method by which these properties can be readily uncovered.

We start with sinusoidal tidal current components u and v; in a general form, they can be written as

$$u = u_0 \cos(\omega t - \varphi_u) \tag{6.73}$$

$$v = v_0 \cos(\omega t - \varphi_v), \tag{6.74}$$

featuring four arbitrary constants u_0, v_0, φ_u and φ_v. Here we consider a fixed location; in general, the four parameters will be functions of x, y, and z.

We write u and v in complex notation using the mathematical identity

$$\cos \alpha = \frac{e^{i\alpha} + e^{-i\alpha}}{2}. \tag{6.75}$$

Hence

$$u = U e^{i\omega t} + U^* e^{-i\omega t} \tag{6.76}$$

$$v = V e^{i\omega t} + V^* e^{-i\omega t}, \tag{6.77}$$

with

$$U = \tfrac{1}{2} u_0 e^{-i\varphi_u}, \qquad V = \tfrac{1}{2} v_0 e^{-i\varphi_v}.$$

The asterisk in (6.76) and (6.77) denotes the *complex conjugate*. Notice that, in spite of the complex notation, the expressions for u in (6.76) and v in (6.77) are still real.

We combine u and v into one complex variable, which we call R:

$$R = u + iv. \tag{6.78}$$

The real part of R is u; its imaginary part, v. The evolution of u and v is thus jointly represented by the time-varying point R in the complex plane.

Combining (6.78) with (6.76) and (6.77) gives

$$R = \overbrace{(U + iV)e^{i\omega t}}^{\text{counterclockwise}} + \overbrace{(U^* + iV^*)e^{-i\omega t}}^{\text{clockwise}}. \tag{6.79}$$

In the complex plane, both time-varying terms describe a *circular* rotation; they are called *phasors* or *rotary components*. Their sense of rotation is opposite: the first is counterclockwise; the second, clockwise. We introduce the shorthand notation for the phasors in (6.79),

$$R_+ = (U + iV)e^{i\omega t}, \qquad R_- = (U^* + iV^*)e^{-i\omega t}, \tag{6.80}$$

so that $R = R_+ + R_-$. The radii of the phasors is $|R_\pm|$.

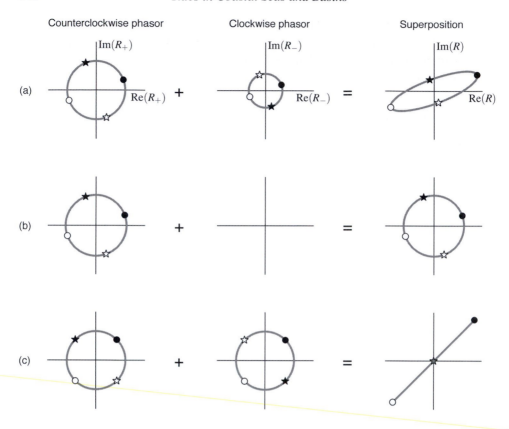

Figure 6.14 The counter-rotating phasors R_+ (left), R_- (center), and their sum (right) for three different shapes of the tidal current field: (a) an ellipse, (b) a circle, (c) rectilinear. Successive phases (i.e., moments in time) are indicated by a black dot, a black star (a quarter period later), an open dot (half a period later), and an open star (three quarters of a period later).

In Figure 6.14, we show the three possible forms. In Figure 6.14a, the superposition of R_+ (left) and R_- (middle), which have different radii, results in an ellipse (right). This is the general case. Following various phases, indicated by dots and stars, explains how the ellipse is produced. At a certain moment, indicated by a black dot, the phasors have identical phases (i.e., the same angle with the horizontal axis). So they add up to create a maximum R. The same happens half a period later (open dot); together these points form the major axis of the ellipse. The length of the semimajor axis is given by the sum of the radii, $|R_+| + |R_-|$, which represents the amplitude of the horizontal tidal current. In between those extremes, the phasors work against each other, which is a consequence of their opposite sense of rotation; this results in a smaller R. In particular, the length of the semiminor axis is given by $|R_+| - |R_-|$. Notice that the polarization of the ellipse follows that of the larger phasor.

Phasors with different radii always result in an ellipse. This leaves us with two special cases. First, an absence of one of the phasors trivially produces a circle for R, as illustrated in Figure 6.14b. Second, phasors of equal radii produce a rectilinear current (Figure 6.14c), since the phasors annul each other in the normal direction, so that the minor axis disappears altogether.

The key properties of the horizontal tidal current (semimajor and semiminor axes, polarization, and shape: ellipse, circle, or rectilinear) thus follow from just two parameters, the radii $|R_{\pm}|$. The one property that remains to be specified is the tilt of the semimajor axis. This angle can be readily obtained by first calculating the angles of R_+ and R_- at an arbitrary moment, $t = 0$ say. They are given by, respectively,

$$\tan \gamma_+ = \frac{\text{Im}(U + iV)}{\text{Re}(U + iV)}, \qquad \tan \gamma_- = \frac{\text{Im}(U^* + iV^*)}{\text{Re}(U^* + iV^*)},$$

where Re and Im indicate the real and imaginary parts. As the phasors turn in opposite directions, their phases must coincide at the mean angle $(\gamma_+ + \gamma_-)/2$, which defines the orientation of the semimajor axis, and hence the tilt.

6.8 Vertical Structure of Tidal Currents: Ekman Dynamics

The tidal current ellipse varies in the vertical. This is already clear from the fact that the current must vanish towards the bottom, which implies a downward shrinking of the semimajor and semiminor axis. An important question is now whether they shrink at the same rate, in the sense that the eccentricity stays constant, or whether the actual character of the ellipse may change in the vertical (e.g., become more rectilinear). The latter turns out to be the case in observations and has been explained theoretically. The necessary ingredients are the Coriolis force and some form of eddy viscosity.

The starting point for analysis is the linear version of the set of momentum equations (6.10) and (6.11), in which we shall represent the stress term as

$$F_x = \rho K \frac{\partial u}{\partial z}, \qquad F_y = \rho K \frac{\partial v}{\partial z},$$

with constant eddy viscosity K. Hence

$$\frac{\partial u}{\partial t} - fv = -g\frac{\partial \zeta}{\partial x} + K\frac{\partial^2 u}{\partial z^2} \tag{6.81}$$

$$\frac{\partial v}{\partial t} + fu = -g\frac{\partial \zeta}{\partial y} + K\frac{\partial^2 v}{\partial z^2}. \tag{6.82}$$

We combine (6.81) and (6.82) by multiplying the latter with i and adding them up. With the earlier definition $R = u + iv$ from (6.78), we can write the resulting equation succinctly as

$$\frac{\partial R}{\partial t} + ifR = -g\Pi + K\frac{\partial^2 R}{\partial z^2}, \tag{6.83}$$

with

$$\Pi = \frac{\partial \zeta}{\partial x} + i\frac{\partial \zeta}{\partial y},$$

which is independent of z. We decompose R into phasors, as in (6.79) and (6.80), and analogously for Π: $\Pi = \Pi_+ + \Pi_-$. Each phasor has to satisfy (6.83), so

$$+i\omega R_+ + ifR_+ = -g\Pi_+ + K\frac{\partial^2 R_+}{\partial z^2} \tag{6.84}$$

$$-i\omega R_- + ifR_- = -g\Pi_- + K\frac{\partial^2 R_-}{\partial z^2}. \tag{6.85}$$

Notice the minus sign in the first term of (6.85), contrary to (6.84). At this point we start to see that the phasors behave differently, which will imply vertical variations in the elliptic properties. We can merge (6.84) and (6.85) into

$$i(f \pm \omega)R_\pm = -g\Pi_\pm + K\frac{\partial^2 R_\pm}{\partial z^2}. \tag{6.86}$$

This second-order ordinary differential equation has constant coefficients and is easy to solve:

$$R_\pm = ig\frac{\Pi_\pm}{f \pm \omega} + A_\pm \cosh(\alpha_\pm z) + B_\pm \sinh(\alpha_\pm z). \tag{6.87}$$

The first term on the right-hand side provides a particular solution, while the second and third terms form the general solution, with arbitrary complex constants $A_\pm$ and $B_\pm$. The hyperbolic functions contain the complex coefficients $\alpha_\pm$, defined by

$$\alpha_\pm = \left(i\frac{f \pm \omega}{K}\right)^{1/2} = (1 \pm i)\left(\frac{\omega \pm f}{2K}\right)^{1/2}.$$

This constant features the *Ekman-layer thickness* $(2K/(\omega \pm f))^{1/2}$. We assume $\omega > f$, in the northern hemisphere. The phasor R_- then has a *thicker* Ekman layer than R_+ (notice that in the southern hemisphere, it would be the other way round). For example, in the North Sea region their ratio is $((\omega + f)/(\omega - f))^{1/2} = 3.3$ for M$_2$.

The constants $A_\pm$ and $B_\pm$ can be solved by applying appropriate boundary condition at the surface and bottom. We impose a free-slip condition at the surface:

$$\frac{\partial R_\pm}{\partial z} = 0 \qquad \text{at } z = 0.$$

Hence $B_\pm = 0$. At the bottom we impose

$$R_\pm = 0 \qquad \text{at } z = -H.$$

Hence

$$A_\pm = -ig \, \frac{\Pi_\pm}{(f \pm \omega)\cosh(\alpha_\pm H)}.$$

The solution is still not fully specified since $\Pi_\pm$ cannot be resolved from the present model. If we prescribe a certain $\Pi_\pm$, then the tidal current ellipses follow from (6.87) at all depth levels. In particular, the solution reveals how the properties of the ellipses vary with depth.

We show an example in Figure 6.15. Here, we prescribed Π via

$$\frac{\partial \zeta}{\partial x} = 5 \times 10^{-6} \cos(\omega t + \pi/2), \qquad \frac{\partial \zeta}{\partial y} = 2 \times 10^{-6} \cos(\omega t).$$

From these expressions, $\Pi_\pm$ are derived in the same way as $R_\pm$ were from u and v, as described in Section 6.7. In Figure 6.15a, we plot the radii of the phasors

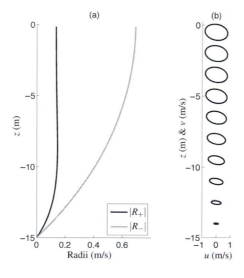

Figure 6.15 Vertical variation of the tidal current ellipse. In (a) we show the radii of the phasors R_+ and R_-, as a function of depth. In (b), we show the tidal ellipses at ten different depths; the center of the ellipse represents the vertical position. The ellipses are drawn in the uv-plane. The parameters are: water depth $H = 15$ m, latitude $\phi = 53°$N, tidal frequency is $\omega = 1.405 \times 10^{-4}$ rad/s (M_2) and eddy viscosity $K = 1 \times 10^{-3}$ m^2/s.

$R_\pm$. For the parameters chosen in this figure (which are representative of M_2 in the North Sea), the Ekman-layer thickness of R_- is about nine meters, that of R_+ about three meters. The effect is clearly seen on the vertical change in the radii: from the surface downward, the radius of phasor R_- starts to decrease almost immediately whereas the radius of R_+ decreases only close to the bottom. In other words, R_- experiences the bottom friction much higher up in the water column than does R_+. Moreover, closer to the bottom, the radii become nearly equal. This is reflected in Figure 6.15b as the tidal current ellipses go from nearly circular at the surface to more rectilinear close to the bottom.

A limitation of the model adopted in this section is the assumption of constant eddy viscosity K. In the case of a vertical stratification in density, vertical eddies are suppressed, implying a reduced eddy viscosity. Such a situation occurs often in coastal areas, notably when a layer of freshwater lies on top of a layer of saline water. The transition between the two layers is called a pycnocline. The reduction of K in the pycnocline will barely affect the phasor R_+, since its Ekman layer is already restricted to the deepest layer. However, the pycnocline may intersect the thicker Ekman layer of R_-, essentially decoupling the upper layer from the frictional layer, which locally enhances its radius. As a result, the properties of the tidal current ellipse in the upper layer will depend on whether the water column is mixed or stratified.

6.9 Tidal Straining

The vertical structure of tidal currents, with stronger currents in the upper layer than near the bottom (see Figures 1.6 and 6.15), has an immediate impact on the vertical distribution of transports. An important example is the movement of fresh and saline water in an estuary. During flood, saline seawater enters, but predominantly in the top layer, thus overlaying the more stagnant deeper and fresher water. This creates a gravitational instability, since the more saline water on top is denser. As a result, the water column will be mixed (Figure 6.16, top panel). During ebb, the reverse occurs: fresher water moves out of the estuary in the top layer, forming a stable vertical stratification (Figure 6.16, lower panel). The result is a tidally induced periodic alternation between mixed and stratified conditions, called SIPS (strain-induced periodic stratification). The "straining" here refers to the horizontal stretching of the isohalines (i.e., contours of equal salinity) by the vertically varying tidal current. Schematically, the process of SIPS is summarized in Figure 6.17.

The degree in which this process occurs, depends primarily on two factors: the strength of the horizontal salinity gradient and the strength of the tidal current. This is expressed in the Simpson number Si, a nondimensional quantity,

$$Si = \frac{g\rho_x H^2}{\rho_* C_D U_T^2}, \tag{6.88}$$

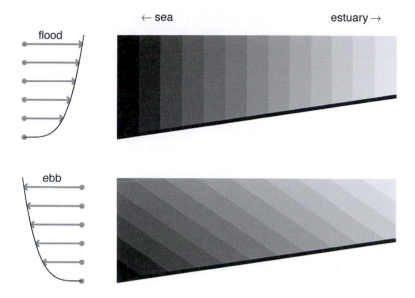

Figure 6.16 The asymmetric effect of flood and ebb on the stratification in an estuary. In this schematic representation, shades of gray represent salinity: from high (black) to low (white). In the upper panel, salinity is vertically mixed; in the lower panel, a vertical gradient in salinity is present.

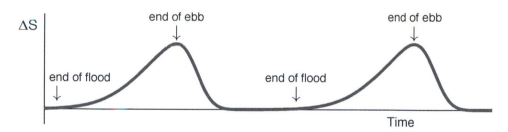

Figure 6.17 Schematic representation of SIPS during two tidal cycles. The vertical stratification is here indicated by ΔS, defined as the difference between bottom and surface salinity. The stratification is built up during ebb, and destroyed during flood.

where U_T is the amplitude of the depth-averaged tidal current, ρ_x the horizontal density gradient, ρ_* a constant reference value of density, H the water depth, and C_D the drag coefficient (see Box 6.1). For strong tidal currents (i.e., small Si), the turbulent kinetic energy in the water column becomes so high that a mixed state occurs even during ebb. For weaker currents (higher Si), one enters the SIPS regime, which may transform to a state of permanent stratification for still weaker currents. The transition between permanently mixed and SIPS lies somewhere around 0.2, but this value is not universal because other factors come into play as well (e.g., wind conditions).

In the previous section we already alluded to the fact that the presence of stratification influences the vertical profile of the tidal current. In the present setting of SIPS, the vertical mixing of momentum is enhanced during flood, when the water column is well mixed. This leads to more vertical uniformity in the current profile. Conversely, during ebb, stratification suppresses vertical eddies, leading to a less uniform profile. The symmetry between flood and ebb in Figure 6.16 is now broken and the profiles become more like those sketched in Figure 6.18. The net effect over a tidal cycle is a residual outflow in the upper layer (where ebb dominates) and a residual inflow in the lower layer (where flood dominates). A similar flow pattern is associated with the classical (gravitational) estuarine circulation, but the underlying causes set them apart. In addition, the asymmetry between ebb an flood profiles implies the presence of overtides (cf. Figure 6.4a). We note that the residual current in the present setting is fundamentally different from the one discussed in Section 6.4, which requires a slope and moreover manifests itself in a depth-averaged sense.

To summarize, the process of SIPS involves an interaction between tidal currents and stratification. At the basis lies a horizontal salinity gradient, which creates a horizontal asymmetry: saline water at one side, freshwater at the other. Combined with the vertical nonuniformity of the tidal current, this can create a periodic alternation of vertically stable and unstable stratification. The resulting alternation between stratified and mixed states is reflected in the vertical tidal profiles and creates an asymmetry between ebb and flood. This does not close the causal loop, because these modified profiles, in turn, affect the straining, etc. This complex interplay of factors cannot be adequately captured in analytical models, so one has to resort to numerical modeling for their quantification.

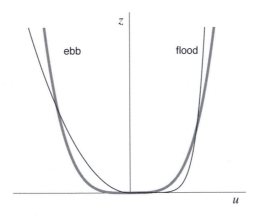

Figure 6.18 Sketch of the vertical tidal current profiles modified by SIPS (black curves), compared to a symmetric tide (gray).

Further Reading

Parker (1991) presents a systematic exploration of the role of the different terms in the generation of shallow-water constituents; an extensive list of them can be found in Le Provost (1991). Ray (2007) examined the propagation of the M_4 overtide in the Atlantic Ocean (cf. Figure 6.3).

For a review on tidally induced residual currents in various settings, see Robinson (1983). In Section 6.4.1, we followed the analysis of the seminal study by Huthnance (1973). An interpretation of tidally induced residual currents in terms of vorticity dynamics can be found in Zimmerman (1981). Particle trajectories under the superposition of tidally induced residual cells and an oscillatory flow, as in (6.37) and (6.38), were first studied by Oonishi and Kunishi (1979). For more background on the mathematical aspects of chaos in dynamical systems, see Nicolas (1995). The potential implication for mixing, in comparison to other mechanisms, was reviewed by Zimmerman (1986). The setting of Section 6.4.3 was explored in more detail by Beerens et al. (1994) with advanced mathematical tools. For an example of numerical modeling of chaotic stirring in a realistic setting, see Ridderinkhof and Zimmerman (1992).

Examples of resonance in tidal basins and straits are discussed in Garrett (1972) and Sutherland et al. (2005); the parameter values mentioned in Section 6.6 were taken from the latter. In Section 6.5 we described a simple example of the "open boundary problem": the notion that the dynamics in the interior of the basin affects the conditions at the open boundary of the model domain, so that the latter cannot be simply regarded as externally prescribed. A more extensive analysis of this problem can be found in Garrett and Greenberg (1977). They also discuss how the tidal signal at the open boundary is affected by engineering works in basins, such as tidal power development. In a numerical modeling study, Shapiro (2011) demonstrated that residual currents can be affected at distances as far as hundred kilometers away from a (hypothetical) tidal stream farm in the Celtic Sea; moreover, the back-effect of turbines on the surrounding flow was shown to substantially reduce the relative gain of adding more of them.

In some sections, we used a linear form of friction: ru/H. For a sinusoidal current with amplitude U_s and frequency ω, $u = U_s \cos \omega t$, the coefficient r can be estimated by making a Fourier expansion of the quadratic friction term $C_D|u|u/H$. The term at frequency ω turns out to be $8C_D U_s u/(3\pi H)$, hence $r = 8C_D U_s/(3\pi)$ (Sutherland et al., 2005). The same expression is obtained by applying an energy argument, requiring that the work done by friction should be equal over a tidal cycle for both forms of friction (Zimmerman, 1982). This comes close to simply taking $r = C_D U_s$; at any rate, the largest uncertainty lies in the estimate of the drag coefficient C_D itself.

The role of eddy viscosity and Ekman dynamics (Section 6.8) is further discussed by Prandle (1982) and Maas and Van Haren (1987), with examples from the North Sea. Visser et al. (1994) examine the effect of stratification on the vertical structure of tidal currents in a ROFI zone (i.e., region of freshwater influence); they show examples of how the ellipticity of surface tidal currents varies depending on the stratification regime.

A simple, empirically based way to describe vertical profiles of tidal currents was proposed by Van Veen (1937): $u = u_0(z + H)^{1/q}$, where in practice q was found to be close to five. We adopted this functional dependence to draw the profiles in Figure 6.18.

Simpson et al. (1990) and Jay and Musiak (1994) are the seminal papers on strain-induced periodic stratification and the consequences for tidal current profiles, respectively, as schematically summarized in Figures 6.16 to 6.18. As noted, a quantitive understanding of the problem requires numerical models. They were employed by Burchard and Hetland (2010) to demonstrate that tidal straining is a dominant factor in the estuarine circulation; in an earlier study, Burchard and Baumert (1998) showed that the ebb-flood asymmetry due to tidal straining (Figure 6.18) is a significant factor in the formation of estuarine turbidity maxima (ETM). Later studies have shown that tidal straining is just one of the mechanisms responsible for temporal variations in eddy viscosity that lead to asymmetries between ebb and flood and net circulation; other factors include, for instance, the presence of M_4 variability or differences in water depth during ebb and flood; see Dijkstra et al. (2017) and references therein.

General overviews on physical processes in estuaries can be found in the book by Officer (1976) and in more recent reviews by Uncles (2002), MacCready and Geyer (2010), and Geyer and MacCready (2014). A topic that falls outside the scope of this introductory textbook is how tides shape their environment; in particular, how tidal currents transport sediment in tidal basins and thus affect their morphology and long-term stability (together with other factors, notably transport by wind-generated waves). Comprehensive overviews on this topic can be found in De Swart and Zimmerman (2009) and Van de Kreeke and Brouwer (2017).

7

Internal Tides

7.1 The Ocean's Inner Unrest

This chapter deals with tides that propagate in the *interior* of oceans and seas – the *internal tides*. They are a special kind of internal wave, namely one at tidal frequency. Their often large vertical movements, sometimes hundreds of meters in the course of a few hours, leave at most a faint trace at the surface. They are thus usually hidden from eyesight. However, they can be easily observed in temperature or current velocity records and are a ubiquitous phenomenon in the world's oceans. Figure 7.1 shows an example: a time record of the vertical movement of an isotherm. Especially during the first five days, the semidiurnal character of the oscillation is unmistakable. The vertical movements extend over heights of as much as 60 m (trough to top).

Internal tides owe their existence to the vertical density stratification in the ocean; they can be regarded as oscillations of the isotherms (or more generally, the isopycnals, the levels of equal density) around their equilibrium depth. Internal tides are not directly generated by the tide-generating potential, but indirectly, via the surface tides discussed in previous chapters. Crucial is the presence of bathymetric forms, which indeed are abundant in the ocean (Figure 5.3). We already saw in Section 5.7 that cross-slope tidal currents exist; they are accompanied by vertical tidal currents over the slope, which periodically move the isopycnals up and down. Like in a stretched rope or string, such an oscillation, forced at one point, engenders waves traveling away from that point, in the form of movements of the isopycnals. Their period reflects the tidal forcing.

The way of propagation of internal tides, and of internal waves in general, is very different from the surface tides or any other kind of surface waves. Because of the continuous nature of the vertical density stratification, internal tides are not confined to horizontal propagation, but propagate also vertically at the same time. This concerns both phase and energy propagation, although (adding another

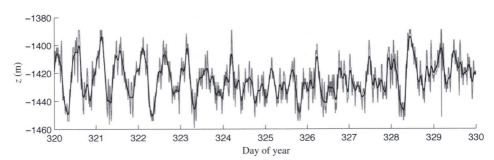

Figure 7.1 Vertical movement of the 6.5°C isotherm around a depth of 1,420 m, during a 10-day period. The original data is shown in gray, with a smoothed version in black to accentuate the overall movement. This record is from the Canary Basin (local water depth: 5,274 m), collected in 2007; it is part of mooring data involving 54 thermistor sensors at different depths, as described by Van Haren and Gostiaux (2009).

element of complexity) they have different directions. One important consequence of the vertical component in internal tidal propagation is that energy supplied to internal tides in the upper layer of the ocean, may at some point enter the abyssal ocean and contribute to local deep mixing.

To distinguish internal tides from the more familiar surface tides, the terms baroclinic and barotropic are often used. In barotropic tides, the levels of equal density (isopycnals) move in unison with those of equal pressure (isobars); both follow the surface movements, with decreasing amplitudes toward the bottom. In contrast, baroclinic tides have large vertical isopycnal movements while in comparison, the isobars stay almost horizontal, being always close to their hydrostatic levels. In other words, the isopycnals are strongly inclined with respect to the isobars.

This chapter is organized as follows. We first discuss at some length the structure of the density stratification in the ocean, as this is a key element in internal tidal propagation. We then derive the equations of motions, restricting ourselves to linear (i.e., small-amplitude) internal tides. Naturally, the equations include the vertical density stratification, which makes them fundamentally different from the equations considered in Chapter 5. We then discuss two methods that are available to study the propagation of internal tides; their applicability depends on the setting (stratification, bathymetry). The next step is to include the forcing by barotropic tides, which allows us to set up a simple model for the generation of internal tides.

7.2 Density Stratification

The temperature distribution in the ocean exhibits a predominantly layered structure. An example is shown in Figure 7.2 for a meridional section in the Pacific

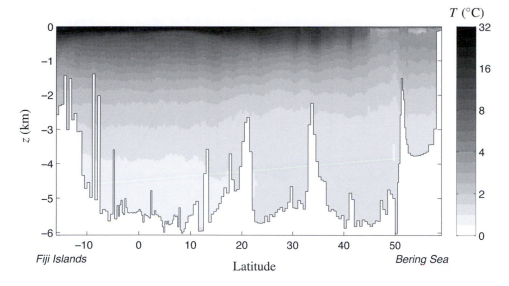

Figure 7.2 Temperature along a transect in the Pacific Ocean (July/August 1993). This transect (WOCE P14) follows the meridian at longitude 179°E, except for an eastward bend in the Bering Sea. It includes the station used in Figure 7.3. Note that the gray scale is logarithmic. Data supplied by the WOCE Hydrographic Program.

Ocean (see also Figure 7.3a for a profile at one location). In a relatively thin sub-surface layer, a few tens of meters thick, the temperature stays nearly constant – the upper mixed layer. This is usually followed by a sharp decline in temperature with increasing depth (the *thermocline*), especially in the tropics and during summer at midlatitudes. The mixed layer and thermocline together occupy just the upper few hundred meters of the water column. Temperature varies relatively little at depths below 1 km, where values are lower than 5°C throughout; the bulk of the ocean is very cold.

The sharp decrease of temperature in the thermocline is reflected in the density,[1] which shows a correspondingly sharp increase with depth (Figure 7.3c). This is followed by a modest but steady increase with depth in the lower part of the water column, which is due to increasing pressure (compressibility). Moreover, density of seawater depends also on salinity (shown in Figure 7.3b), so we can write

$$\rho = \rho(p, T, S), \tag{7.1}$$

---

[1] Throughout this chapter, density always refers to the actual, *in-situ* density, in contrast to the so-called *potential* density, which is the virtual density a water parcel would have if it were brought adiabatically to some reference level.

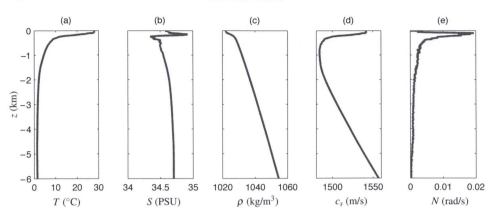

Figure 7.3 Measured profiles of (a) temperature T and (b) salinity S in the Pacific Ocean (location 12.5°N, 179°E). From these profiles, (c) density ρ, (d) the speed of sound c_s, and (e) the buoyancy frequency N were calculated. Some smoothing was applied to obtain the profile of N (in measurements of salinity, notably, even variations within the margin of accuracy already cause perceptible noisiness in the derivative $d\rho/dz$, and hence in N). Data in (a) and (b) supplied by the WOCE Hydrographic Program.

the *equation of state* for seawater. The function is a complicated empirically based polynomial expression (for which publicly available standard routines exist). An equation of state fundamentally refers to a state of thermodynamic equilibrium, i.e., an isothermal state in hydrostatic balance.[2] Plainly, the actual state of the ocean, considered in its entirety, is far removed from thermodynamic equilibrium. Still, the notion of thermodynamic equilibrium can be applied, but in a *local* sense (following the so-called local equilibrium assumption). Accordingly, equations of state are formulated in terms of intensive quantities like density, which are defined locally.

The steady increase of density with depth in Figure 7.3c seems to suggest gravitational stability. However, we need to consider the problem more carefully. Local gravitational stability refers to the response of a water parcel when it is given a small vertical displacement: if it tends to move backward, the water column is gravitationally stable. Now, if a water parcel is moved upward, the point is not whether its original density is higher than that of its surroundings, but whether this holds for its actual density. In other words, we have to take into account how the density of a vertically moving water parcel changes. For this, we need to specify the movement from a thermodynamic point of view. We shall assume that the motion of water parcels is adiabatic and reversible; this means that they do not exchange

---

[2] In addition, thermodynamic equilibrium requires a uniform relative chemical potential, which would imply a strong increase in salinity with depth due to the presence of gravity; in this respect, too, the ocean as a whole is evidently far removed from thermodynamic equilibrium.

heat or salinity with their surroundings, and that their movement is sufficiently slow for the water parcels to be always in a state of (local) thermodynamic equilibrium.

Based on this assumption, gravitational stability is commonly expressed in terms of a quantity N^2, where N is called the *buoyancy frequency* (or Brunt–Väisälä frequency). If $N^2 > 0$ the water column is locally gravitationally stable; if $N^2 < 0$, it is unstable. The expression for N^2, which will be derived in Section 7.3.2, reads

$$N^2 = -\frac{g}{\rho}\left(\frac{d\rho}{dz} + \frac{\rho g}{c_s^2}\right). \tag{7.2}$$

As expected, the right-hand side features the vertical density gradient $d\rho/dz$. Notice that it is negative when density increases with depth, which gives a positive contribution to N^2. The right-hand side contains a second term, which involves the speed of sound c_s. Like density, c_s is a state variable that can be calculated from an equation of state for given p, T, and S. A typical profile of c_s is shown in Figure 7.3d. In the upper layer, c_s decreases with depth due to the sharp drop in temperature; in the lower part, where T (and S) are fairly uniform, the increase of c_s with depth is due to increasing pressure. The second term on the right-hand side of (7.2) has a destabilizing effect (i.e., it reduces N^2). In the deeper parts of the ocean, the two terms on the right-hand side are of the same order of magnitude, but of opposite signs; the result is a mostly marginally stable state.

We show a profile of N in Figure 7.3e. The thermocline[3] stands out as a strong peak in N, while in the deeper parts of the water column, N becomes very small. This is shown more generally in Figure 7.4 for a transect in the Pacific Ocean, which confirms that N varies greatly: from the order of 10^{-4} in the deepest layers to 10^{-2} rad/s in the thermocline.

Gravitational stability is a measure of how strongly gravity can act as a restoring force in the interior of the ocean. Internal tides owe their existence to this restoring force (in combination with the Coriolis force, discussed later); after all, this is what allows the oscillations to happen. Buoyancy frequency N reflects the strength of gravitational stability. It has the unit of frequency, which invites a comparison with the tidal frequency ω. We see immediately from the values in Figure 7.4 that N nearly everywhere exceeds the frequencies of semidiurnal and diurnal tides, with the possible exception of pockets of weak (or unstable) stratification in the deepest parts of the ocean, or in the upper mixed layer. So, we can generally assume that $\omega < N$. This inequality can be supplemented by the one involving the Coriolis parameter, inferred from Figure 5.4. The latter depends on latitude. As a result,

---

[3] In this chapter, we refer to the subsurface peak in N as the thermocline, as it is primarily due to the strong vertical variation in temperature. A more general term is *pycnocline*, which refers to a sharp change in density, either due to temperature or salinity – the latter is dominant in fjords, for example.

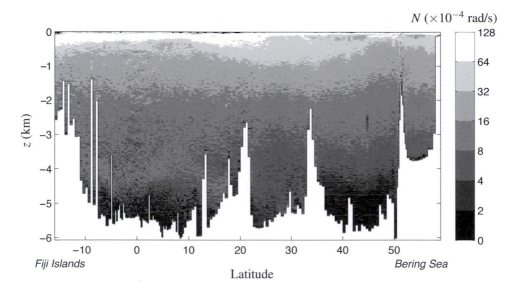

N ($\times 10^{-4}$ rad/s)

Figure 7.4 As in Figure 7.2, but now showing the buoyancy frequency N. Note again that the gray scale is logarithmic.

at subpolar latitudes, we generally have the following combined inequality for semidiurnal tidal constituents

$$|f| < \omega < N. \tag{7.3}$$

In internal-wave theory, f and N act as key parameters, as we shall see in later sections.

Finally, we consider a couple of simple representations of the vertical stratification of density. In numerical models, one can just use empirical profiles of $N(z)$ without a need for further simplification. In theoretical studies, N is often taken constant, which is motivated more by mathematical expedience than by an aspiration to realism (recall that N varies by two orders of magnitude in the ocean). Still, much can be learned from those models, but caution is needed when it comes to the significance of the results for the actual ocean.

We can come somewhat closer to reality by constructing a piecewise constant profile of N, involving three layers, as sketched in Figure 7.5. (For a reason that will soon become clear, we plot N^2 rather than N in this figure.) At the top, there is a mixed layer of thickness d; here we simply take $N = 0$. Below the mixed layer, we have a thermocline of thickness ε and strength $N^2 = g'/\varepsilon$, with constants g' and ε. Notice that in this construction, the *area* enclosed by the thermocline equals g' in the N^2 versus z graph. This offers an easy way to estimate a suitable value of g' from an empirical profile of $N^2(z)$. Finally, the deeper part of the ocean is represented by constant buoyancy frequency N_c.

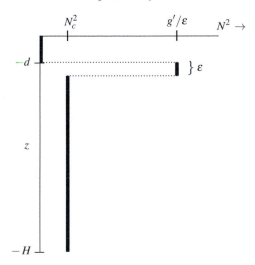

Figure 7.5 Schematic representation of the ocean's stratification in terms of N^2, featuring an upper mixed layer, a thermocline, and a weakly stratified deep layer.

The profile of Figure 7.5 can be simplified by taking the limit $\varepsilon \to 0$. The thermocline then becomes infinitely strong ($g'/\varepsilon \to \infty$) and vanishingly thin. It effectively becomes an interface across which the density has a discontinuity (hence the infinite N). This is not unlike the upper surface of the ocean, where we have a jump from the density of air to the density of water, but in the case of the interfacial thermocline, the magnitude of the jump is much smaller. This is expressed by the effective gravity g', which is typically three orders of magnitude smaller than g.

7.3 Equations of Motion

We already discussed the fundamental equations of motion in Section 5.2, but one assumption must now obviously be abandoned: we can longer suppose density ρ to be constant. Variations in density enter the problem in two steps. We start with a static equilibrium state, denoted with index "0"; this state is described by the hydrostatic balance (5.12),

$$\frac{dp_0}{dz} = -\rho_0 g, \tag{7.4}$$

where ρ_0 is a function of z. We regard internal waves as perturbations of this state. Accordingly, we write pressure and density as superpositions of their static values and the perturbations (the latter denoted by a prime):

$$p = p_0(z) + p'(t,x,y,z) \tag{7.5}$$

$$\rho = \rho_0(z) + \rho'(t,x,y,z). \tag{7.6}$$

In a way, these are empty statements, since we can always write p and ρ like this as long as p' and ρ' have not been specified. It is only in later approximations that the split becomes consequential.

We consider small-amplitude waves in this chapter, which means that we can neglect the advective terms in the momentum equations. Moreover, we adopt the f-plane and neglect friction. Thus, we can take the momentum equations (5.9) to (5.11) as a starting point,

$$\rho\left(\frac{\partial u}{\partial t} - fv\right) = -\frac{\partial p}{\partial x} \tag{7.7}$$

$$\rho\left(\frac{\partial v}{\partial t} + fu\right) = -\frac{\partial p}{\partial y} \tag{7.8}$$

$$\rho\frac{\partial w}{\partial t} = -\frac{\partial p}{\partial z} - \rho g. \tag{7.9}$$

In the present setting, however, the equations are not yet entirely linear, since the left-hand sides implicitly include products of perturbations, e.g., $\rho' \partial u / \partial t$. Neglecting these terms as well, we obtain the truly linear set

$$\rho_0\left(\frac{\partial u}{\partial t} - fv\right) = -\frac{\partial p'}{\partial x} \tag{7.10}$$

$$\rho_0\left(\frac{\partial v}{\partial t} + fu\right) = -\frac{\partial p'}{\partial y} \tag{7.11}$$

$$\rho_0\frac{\partial w}{\partial t} = -\frac{\partial p'}{\partial z} - \rho' g. \tag{7.12}$$

In the step from (7.9) to (7.12), we also used the hydrostatic balance (7.4) to remove the static part from the right-hand side. On the right-hand sides of (7.7) and (7.8), derivatives of p_0 are zero, hence the remaining terms are derivatives of p' in (7.10) and (7.11).

7.3.1 Bousssinesq Approximation

In the ocean, density varies by at most a few percentage points (see Figure 7.3c). We may exploit this fact to further simplify (7.10) to (7.12), but, as we shall see, the paradoxical outcome is that variations in density turn out to be essential in one respect, and negligible in another.

At the heart of the problem lies the distinction between inertial and gravitational mass, known from classical mechanics.[4] Inertial mass is a measure of how strongly

---

[4] It was shown observationally by Eötvös that the two can be identified numerically; later, conceptual equality was implied by Einstein's theory of general relativity.

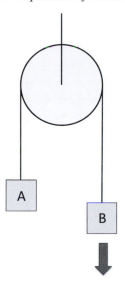

Figure 7.6 Atwood's machine: two bodies are connected by a string passing over a pulley. Here, body A has mass m and body B has a slightly larger mass, $m + \Delta m$.

a body resists a change in velocity when subjected to a force; gravitational mass determines the strength of the force by which a body is attracted in a field of gravity. The former has a passive connotation, the latter an active one.

It is instructive to consider a simple mechanical example: Atwood's machine (Figure 7.6). We assume that body B has a slightly larger mass $(m + \Delta m)$ than body A (m), with $\Delta m \ll m$. The net force of gravity working on the connected bodies is $g\Delta m$, which exerts a pull in the direction of the arrow. This net force acts on the total mass of bodies A and B, $2m + \Delta m$. Hence the acceleration a is

$$a = \frac{g\Delta m}{2m + \Delta m}.$$

Since $\Delta m \ll m$, we can approximate this expression by

$$a \approx \frac{g\Delta m}{2m}.$$

This amounts to neglecting the difference in mass so far as the *inertial* mass is concerned; in the *gravitational* mass (in the numerator) the difference remains essential since otherwise their would be no net force.

In our setting, mass is substituted by density; we can similarly make the distinction between inertial density, represented by the coefficient ρ_0 on the left-hand sides of (7.10) to (7.12), and gravitational density, represented by the term $\rho'g$ on the right-hand side of (7.12). By analogy with the example of Atwood's machine, we ignore variations in density so far as inertial density is concerned, i.e., we replace $\rho_0(z)$ by a constant reference density ρ_*, but we leave the gravitational-density term unaltered. Hence (7.10) to (7.12) become

$$\rho_*\left(\frac{\partial u}{\partial t} - fv\right) = -\frac{\partial p'}{\partial x} \tag{7.13}$$

$$\rho_*\left(\frac{\partial v}{\partial t} + fu\right) = -\frac{\partial p'}{\partial y} \tag{7.14}$$

$$\rho_*\frac{\partial w}{\partial t} = -\frac{\partial p'}{\partial z} - \rho'g. \tag{7.15}$$

This simplification is mathematically convenient, for we have now obtained a set of equations with constant coefficients. By lack of a better name, we refer to this as the "Boussinesq approximation," but we note that the term is sometimes used in a broader sense in the literature (i.e., beyond the momentum equations).

7.3.2 Thermodynamic Energy Principles

In the equation of state (7.1), we have taken pressure p, temperature T and salinity S as the three independent state variables. This is natural from an observational point of view, as these variables can be easily measured.

However, at this point it is convenient to use entropy η instead of T as one of the independent state variables, for reasons that will become clear shortly. We thus express density as $\rho(p, \eta, S)$, for a certain function that we can leave unspecified here. Material changes in density, i.e. following a parcel, can then be expressed as

$$\frac{D\rho}{Dt} = \left(\frac{\partial \rho}{\partial p}\right)_{\eta S}\frac{Dp}{Dt} + \left(\frac{\partial \rho}{\partial \eta}\right)_{pS}\frac{D\eta}{Dt} + \left(\frac{\partial \rho}{\partial S}\right)_{p\eta}\frac{DS}{Dt}, \tag{7.16}$$

with the material derivative defined in (5.17). The subscripts on the right-hand side of (7.16) are added to indicate that these variables are kept constant in the respective partial derivatives. The first partial derivative on the right-hand side has a special significance, for it is related to the speed of sound c_s, as[5]

$$\frac{1}{c_s^2} = \left(\frac{\partial \rho}{\partial p}\right)_{\eta S}. \tag{7.17}$$

---

[5] This can be seen by deriving a wave equation in the absence of gravity and the Coriolis force ($g = 0$, $f = 0$), which leaves only pressure gradients as a restoring force for wave propagation. This results in a wave equation where c_s features as the phase speed, see, e.g., Vallis (2006, §1.8).

As already mentioned in Section 7.2, we assume that water parcels do not exchange heat or salinity with their surroundings as they move around. This means that the parcel's entropy and salinity are conserved:

$$\frac{D\eta}{Dt} = 0, \qquad \frac{DS}{Dt} = 0. \tag{7.18}$$

Hence the last two terms on the right-hand side of (7.16) vanish identically, and the only remaining terms are

$$\frac{D\rho}{Dt} = \frac{1}{c_s^2} \frac{Dp}{Dt}. \tag{7.19}$$

We split pressure and density according to (7.5) and (7.6) and neglect all the products of perturbation terms, in accordance with the assumption of small-amplitude waves,

$$\frac{\partial \rho'}{\partial t} + w \frac{d\rho_0}{dz} = \frac{1}{c_s^2} \left(\frac{\partial p'}{\partial t} + w \frac{dp_0}{dz} \right). \tag{7.20}$$

Using the hydrostatic balance (7.4), we can rewrite this as

$$\frac{\partial \rho'}{\partial t} + w \left(\frac{d\rho_0}{dz} + \frac{\rho_0 g}{c_s^2} \right) = \frac{1}{c_s^2} \frac{\partial p'}{\partial t}. \tag{7.21}$$

We compare the expected magnitude of the term on the right-hand side with the first term on the left-hand side. Assuming a primary balance between the two terms on the right-hand side of (7.15), we can express the scale of p' as $[p'] = [\rho']gH$. This means that the ratio of the two terms is gH/c_s^2, which is much smaller than one (see Figure 7.3d for typical values of c_s). Hence the primary balance must be between the terms on the left-hand side of (7.21) and the term on the right-hand side can be neglected:

$$\frac{\partial \rho'}{\partial t} + w \left(\frac{d\rho_0}{dz} + \frac{\rho_0 g}{c_s^2} \right) = 0. \tag{7.22}$$

In the expression in brackets we already see the appearance of the buoyancy frequency N, as in (7.2), but now with specific reference to the static background state ρ_0. If the factor in brackets is negative, an upward vertical velocity ($w > 0$) will locally increase the density (ρ') with respect to the background state, which means that the fluid is stably stratified. This confirms the classification concerning N^2 as set out in Section 7.2, according to which $N^2 > 0$ characterizes a stably stratified fluid.

Finally, we emphasize that thermodynamic energy principles were pivotal in the step from (7.16) to (7.19), via (7.18). The resulting equation (7.22) is an expression of those principles, even though it is cast in terms of density (and not energy as such).

7.3.3 Mass Conservation

The equation for mass conservation is again (5.5):

$$\frac{\partial \rho}{\partial t} + \frac{\partial}{\partial x}(\rho u) + \frac{\partial}{\partial y}(\rho v) + \frac{\partial}{\partial z}(\rho w) = 0. \tag{7.23}$$

Splitting density according to (7.6) and neglecting all products of perturbation terms results in

$$\frac{\partial \rho'}{\partial t} + \rho_0 \frac{\partial u}{\partial x} + \rho_0 \frac{\partial v}{\partial y} + \rho_0 \frac{\partial w}{\partial z} + w \frac{d\rho_0}{dz} = 0. \tag{7.24}$$

Combined with (7.22) we obtain

$$\frac{\partial u}{\partial x} + \frac{\partial v}{\partial y} + \frac{\partial w}{\partial z} = \frac{wg}{c_s^2}. \tag{7.25}$$

If we assume that $\partial w / \partial z$ scales as $[w]/H$, we see that the term on the right-hand side is very small in comparison, since $gH/c_s^2 \ll 1$. Thus, we can suppose that the primary balance is between the terms on the left-hand side,

$$\frac{\partial u}{\partial x} + \frac{\partial v}{\partial y} + \frac{\partial w}{\partial z} = 0. \tag{7.26}$$

7.3.4 Single Governing Equation

The full set of equations derived so far consists of the momentum equations (7.13), (7.14), and (7.15), the thermodynamic equation (7.22), and the continuity equation (7.26). Together, they provide five equations for the five unknowns u, v, w, p', and ρ'. In addition, equations of state are needed to determine ρ_0 and c_s (and hence N) from temperature and salinity profiles; hereafter we will regard N as a given function of z. We replace perturbation density ρ' with *buoyancy* b,

$$b = -g\rho'/\rho_*. \tag{7.27}$$

The aforementioned set of equations then becomes

$$\frac{\partial u}{\partial t} - fv = -\frac{1}{\rho_*}\frac{\partial p'}{\partial x} \tag{7.28}$$

$$\frac{\partial v}{\partial t} + fu = -\frac{1}{\rho_*}\frac{\partial p'}{\partial y} \tag{7.29}$$

$$\frac{\partial w}{\partial t} = -\frac{1}{\rho_*}\frac{\partial p'}{\partial z} + b \tag{7.30}$$

$$\frac{\partial b}{\partial t} + N^2 w = 0 \tag{7.31}$$

$$\frac{\partial u}{\partial x} + \frac{\partial v}{\partial y} + \frac{\partial w}{\partial z} = 0. \tag{7.32}$$

Due to the Boussinesq approximation, and the corresponding definition of b, the buoyancy frequency N has undergone a small (inconsequential) change in its factor, in comparison with (7.2), for in (7.31) it has become

$$N^2 = -\frac{g}{\rho_*}\left(\frac{d\rho_0}{dz} + \frac{\rho_0 g}{c_s^2}\right). \tag{7.33}$$

The set (7.28)–(7.32) can be reduced to one equation for the vertical velocity w. We choose this variable because the boundary conditions at the bottom and surface are most easily posed in terms of w. The derivation involves a number of steps. First, taking $\partial/\partial z$ of (7.29), and $\partial/\partial y$ of (7.30), and subtracting the results, gives

$$\frac{\partial}{\partial t}\left(\frac{\partial w}{\partial y} - \frac{\partial v}{\partial z}\right) = f\frac{\partial u}{\partial z} + \frac{\partial b}{\partial y}. \tag{7.34}$$

Similarly, taking $\partial/\partial z$ of (7.28), and $\partial/\partial x$ of (7.30), gives

$$\frac{\partial}{\partial t}\left(\frac{\partial u}{\partial z} - \frac{\partial w}{\partial x}\right) = f\frac{\partial v}{\partial z} - \frac{\partial b}{\partial x}. \tag{7.35}$$

Finally, combining $\partial/\partial y$ of (7.28) and $\partial/\partial x$ of (7.29):

$$\frac{\partial}{\partial t}\left(\frac{\partial v}{\partial x} - \frac{\partial u}{\partial y}\right) = f\frac{\partial w}{\partial z}, \tag{7.36}$$

where we used the continuity equation (7.32) to simplify the term with f. The left-hand sides of (7.34), (7.35), and (7.36) describe changes in vorticity, $\nabla \times \vec{u}$.

Subtracting $\partial^2/\partial y\partial t$ of (7.34) and $\partial^2/\partial x\partial t$ of (7.35), gives

$$\frac{\partial^2}{\partial t^2}\left(\nabla_h^2 w - \frac{\partial}{\partial z}\left[\frac{\partial u}{\partial x} + \frac{\partial v}{\partial y}\right]\right) + f\frac{\partial^2}{\partial t\partial z}\left(\frac{\partial v}{\partial x} - \frac{\partial u}{\partial y}\right) - \nabla_h^2\frac{\partial b}{\partial t} = 0,$$

with

$$\nabla_h^2 = \frac{\partial^2}{\partial x^2} + \frac{\partial^2}{\partial y^2}.$$

The term in square brackets can be rewritten with (7.32); the Coriolis term, with (7.36); and the last term, with (7.31). The result is an equation for w alone:

$$\boxed{\frac{\partial^2}{\partial t^2}\nabla^2 w + f^2\frac{\partial^2 w}{\partial z^2} + N^2\nabla_h^2 w = 0,} \tag{7.37}$$

with

$$\nabla^2 = \frac{\partial^2}{\partial x^2} + \frac{\partial^2}{\partial y^2} + \frac{\partial^2}{\partial z^2}.$$

We shall consider waves of the form

$$w = \hat{w}(x, y, z) \exp(-i\omega t), \tag{7.38}$$

for a certain function $\hat{w}$ (which is to be determined) and a given tidal frequency ω. Here, the real part is implied. Substitution in (7.37) gives[6]

$$(f^2 - \omega^2)\frac{\partial^2 \hat{w}}{\partial z^2} + (N^2 - \omega^2)\nabla_h^2 \hat{w} = 0. \tag{7.39}$$

The horizontal coordinates x and y appear in a completely symmetric way; so, without loss of generality, we may select wave propagation in any horizontal direction. We shall consider plane waves propagating in the x direction, which are uniform in y, hence

$$(f^2 - \omega^2)\frac{\partial^2 \hat{w}}{\partial z^2} + (N^2 - \omega^2)\frac{\partial^2 \hat{w}}{\partial x^2} = 0. \tag{7.40}$$

There are now two ways to proceed. One is to introduce the horizontal propagation explicitly as $\sim \exp(ikx)$, which reduces the problem to finding the vertical structure of $\hat{w}$. This solution, in terms of *vertical modes*, will be pursued in Section 7.4. At the same time, (7.40) exhibits a remarkable similarity in x and z, so it seems natural to exploit this by treating them alike. This can be done by introducing new, so-called *characteristic coordinates*; they bring out the diagonal nature of the wave propagation in the xz-plane (Section 7.5).

These seemingly disparate ways of dealing with (7.40) represent, in fact, two different angles of looking at the same phenomenon. However, each method has its own strength and limitation. In the method of vertical modes, the problem is conceived as *separable* in the x and z coordinates, i.e., each vertical mode is written as a product of a function of x times a function of z, and the modes propagate independently of each other. The bottom has to be flat for this solution to be exact, lest the modes become coupled. On the other hand, $N(z)$ may take any form. For the method of characteristics, it is essentially the other way round. No separation

---

[6] At this point it is worth while to note that we have *not* used the hydrostatic approximation in this chapter. If we would redo the derivation under the hydrostatic approximation, i.e., assuming $\partial p'/\partial z = -\rho' g$ instead of (7.15), the resulting equation would have the same structure as (7.39) except that the factor $N^2 - \omega^2$ is replaced by N^2. This means that the hydrostatic approximation implicitly involves the assumption $N \gg \omega$. Plainly, adopting the hydrostatic approximation would not make the problem fundamentally simpler; it merely changes a coefficient.

of variables is required, so the bottom may take any shape. However, for x and z to play similar roles in (7.40), N has to be constant.[7]

The two methods are equally applicable in a system with constant N and a flat bottom and then lead to the same solution. In contrast, neither is even approximately applicable in a system with strongly varying N over steep slopes. It so happens that precisely these conditions are of central importance in the ocean, since this is where much of the internal tidal generation takes place: over the higher part of the continental slope near the thermocline. To study these processes, one has to resort to numerical modeling. Still, the methods to be discussed in the following sections are valuable tools to get insight into the fundamentals of internal tidal propagation.

7.4 Vertical Modes

In this section, we consider a configuration that is bounded by a flat bottom at $z = -H$ and a surface at $z = 0$. We regard the latter as a rigid-lid, at which we impose $w = 0$. The motivation for this approximation is that vertical movements in the interior, being of the order of tens or even a few hundreds of meters, are very much larger than the surface movements. In comparison, the latter can thus be conceived as immovable. Notice that the rigid-lid experiences oscillations in pressure from the fluid underneath and, by Newton's third law, exerts an equal but opposite pressure on the fluid.

We seek solutions of the form

$$\hat{w}(x, z) = W(z) \exp ikx. \tag{7.41}$$

Substitution in (7.40) results in an ordinary differential equation for W:

$$\boxed{W'' + k^2 \frac{N^2(z) - \omega^2}{\omega^2 - f^2} W = 0,} \tag{7.42}$$

where primes denote derivatives to z. We pose the boundary conditions at the surface (regarded as a rigid-lid) and the horizontal bottom as

$$W = 0 \quad \text{at } z = -H, 0. \tag{7.43}$$

Together, (7.42) and (7.43) form an eigenvalue problem, which for given ω, f and $N(z)$ has an infinite number of solutions W_n (eigenfunctions, vertical modes) with corresponding wavenumbers k_n.

---

[7] There are a few special profiles of N that also allow for a solution in terms of characteristics; see Krauss (1966, §151).

Without proof, we state an important mathematical property of (7.42) and (7.43), namely the orthogonality of its eigenfunctions. Specifically, for any W_n and W_j, with $n \neq j$, the following identities hold

$$\int_{-H}^{0} dz\, W_n'' W_j = 0, \qquad \int_{-H}^{0} dz\, \frac{N^2(z) - \omega^2}{\omega^2 - f^2}\, W_n W_j = 0. \qquad (7.44)$$

Before examining specific solutions, we derive some more general properties that follow from (7.42) and (7.43). The first one concerns the horizontal currents. Assuming a similar form as for w, i.e.,

$$(u, v) = (U, V) \exp i(kx - \omega t), \qquad (7.45)$$

with U and V functions of z, we find their expressions from (7.32) and (7.36), respectively,

$$U = \frac{i}{k} W', \qquad V = \frac{f}{\omega k} W'. \qquad (7.46)$$

The component V owes its presence to the Coriolis force. The boundary conditions (7.43) imply that the depth-integrated horizontal velocities vanish, i.e.,

$$\int_{-H}^{0} dz\, U = 0, \qquad \int_{-H}^{0} dz\, V = 0. \qquad (7.47)$$

This is one of the characteristic properties of internal tides (or, internal waves in general), in clear contrast with surface tides.

For convenience, we introduce a short-hand notation for the coefficient in (7.42),

$$m^2 = k^2 \frac{N^2(z) - \omega^2}{\omega^2 - f^2}. \qquad (7.48)$$

Notice that m depends on z. Solutions to (7.42) may exhibit two kinds of behavior, depending on the sign of m^2. Firstly, oscillatory in those parts of the water column where m is real, i.e., $m^2(z) > 0$. For this to be possible, one of the two following inequalities must hold throughout this part of the water column:

$$\boxed{\text{(I)} \quad N(z) < \omega < |f| \quad \text{or} \quad \text{(II)} \quad |f| < \omega < N(z).} \qquad (7.49)$$

This puts a constraint on the allowable wave frequencies. The prevailing situation in the ocean is $N > |f|$, and for semidiurnal tides condition (II) is mostly satisfied, as noted empirically in (7.3). In layers where neither of the conditions (7.49) is satisfied, the solution is quasi-exponentially decaying in z, since $m^2 < 0$. For internal tides of a given frequency ω to exist at all, one of the inequalities in (7.49) should be satisfied in at least part of the water column.

7.4.1 Uniform Stratification

We now solve (7.42) and (7.43) for the simplest case, choosing a constant N. Supposing that one of the conditions in (7.49) is satisfied, we can state the general solution of (7.42) as

$$W = A \sin mz + B \cos mz, \tag{7.50}$$

for certain constants A and B. From the boundary conditions (7.43) it is obvious that the cosine term must vanish, but we take a more formal route to elucidate the general procedure. We collect the boundary conditions in a matrix:

$$\begin{pmatrix} 0 & 1 \\ -\sin mH & \cos mH \end{pmatrix} \begin{pmatrix} A \\ B \end{pmatrix} = \begin{pmatrix} 0 \\ 0 \end{pmatrix}. \tag{7.51}$$

To have nontrivial solutions for the pair (A, B), the determinant of the matrix in (7.51) must be zero. This solvability condition leads to the *dispersion relation*. Specifically, the determinant is zero if $\sin mH = 0$, hence

$$m_n = \pm \frac{n\pi}{H}, \qquad \text{for} \quad n = 1, 2, 3, \cdots. \tag{7.52}$$

Via the definition of m in (7.48), we obtain the dispersion relation, which relates wavenumber k and frequency ω,

$$k_n = \pm \frac{n\pi}{H} \left(\frac{\omega^2 - f^2}{N^2 - \omega^2} \right)^{1/2}, \qquad n = 1, 2, 3, \cdots. \tag{7.53}$$

For a given wave frequency ω, one thus finds an infinite set of wavenumbers k_n. Waves become shorter, i.e. $|k_n|$ increases, with increasing mode number n. Furthermore, for $\omega \to |f|$, we find the long-wave limit $|k_n| \to 0$; and for $\omega \to N$, the shortwave limit $|k_n| \to \infty$, as illustrated in Figure 7.7.

We also see from (7.53) that the second mode is twice as short as the first one; the third three times as short, etc. In other words, the wavenumbers are commensurable:

$$k_n / k_l = n / l. \tag{7.54}$$

As a consequence, a superposition of modes must be horizontally periodic. We note that this property does not hold for general $N(z)$; it is a peculiarity of the case of constant N.

Alternatively, we can rewrite (7.53) to express the wave frequency as a function of k_n and m_n:

$$\omega^2 = \frac{N^2 k_n^2 + f^2 m_n^2}{k_n^2 + m_n^2}. \tag{7.55}$$

From (7.55) we derive the horizontal phase speed $c = \omega/k_n$ and group velocity $c_g = \partial\omega/\partial k_n$,

$$c = \pm \frac{H\omega}{n\pi} \left(\frac{N^2 - \omega^2}{\omega^2 - f^2}\right)^{1/2} \tag{7.56}$$

$$c_g = \pm \frac{H}{n\pi} \frac{[(\omega^2 - f^2)(N^2 - \omega^2)^3]^{1/2}}{\omega(N^2 - f^2)}. \tag{7.57}$$

A plus (minus) sign in (7.53) corresponds to a plus (minus) sign in (7.56) and (7.57). The group velocity c_g indicates how fast the wave energy travels; the phase speed c, how fast the wave's crests and troughs travel. The former determines where the energy goes and where the internal waves manifest themselves. Notice that c and c_g are in the same direction if $N > |f|$, but horizontally opposed if $N < |f|$. The phase speed and group velocity both are inversely proportional to mode number n: higher modes propagate more slowly. Their behavior as a function of wave frequency is sketched in Figure 7.7.

Returning to (7.51), we see that B must be zero, while A may take any (complex) value. Thus, the vertical modes are

$$W_n = A_n \sin\left(\frac{n\pi z}{H}\right), \qquad n = 1, 2, 3, \cdots. \tag{7.58}$$

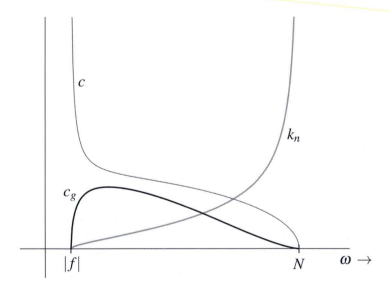

Figure 7.7 Qualitative behavior of wavenumber k, phase speed c, and group velocity c_g as a function of wave frequency ω, for a fixed mode number. They follow from (7.53), (7.56), and (7.57), respectively. Note that the curve for k comes with its own scale, as its unit is different from c and c_g. Here we assume $N > |f|$.

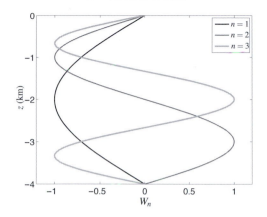

Figure 7.8 The first three vertical modes W_n, from (7.58), with $A_n = 1$ m/s and for $H = 4$ km.

Their structure is independent of the wave frequency ω. This is another peculiarity of the case of constant N; for general profiles $N(z)$, the modes take different structures for different frequencies (see the following section). The first three vertical modes of (7.58) are shown in Figure 7.8.

Returning to the original equation (7.37) for w, we notice its linear character, which implies that we can add up modes; any superposition automatically satisfies the equation. By combining the time-dependence from (7.38), the x-dependence from (7.41), and the model structure from (7.58), we construct the superposition

$$w = \sum_n A_n \sin\left(\frac{n\pi z}{H}\right) \cos(k_n x - \omega t), \qquad (7.59)$$

where we took the real part (assuming A_n to be real). We select positive k_n, i.e., rightward-propagating waves. To obtain the most general solution, we should allow A_n to be complex in (7.58), and include waves propagating in the opposite direction. However for the present purposes, expression (7.59), as it stands, is adequate.

With (7.46), we already showed how the fields u and v can be obtained. Similarly, expressions for p' and b can be derived from (7.28) and (7.31). Apart from a coefficient, p' has essentially the same structure as u. Moreover, isopycnal displacements ζ can be calculated from $w = \partial\zeta/\partial t$. For the first three modes, the spatial structures of all these fields are shown in Figure 7.9, as snapshots at $t = 0$. Notice in particular the change(s) in sign in u and v in the vertical, in accordance with (7.47). With increasing mode number, the horizontal and vertical scales become smaller (i.e., larger k_n and m_n). In time, the pattern in each panel would move to the right, but more slowly so for higher mode numbers, as indicated by (7.56) and (7.57).

In any realistic situation, the internal tidal signal consists of a superposition of modes. In Figure 7.10, we show examples of superpositions for horizontal current

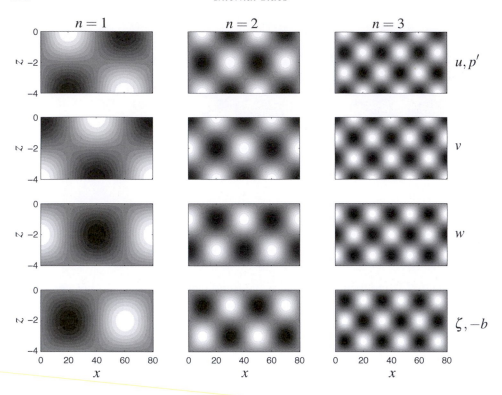

Figure 7.9 Spatial structure of the first three modes, plotted individually for rightward propagating waves, shown as a snapshot at $t = 0$. Upper panels show the horizontal velocity component u; the second row, the transverse velocity v; the third row, the vertical velocity w; and the lowest panels the isopycnal displacement ζ. The first and fourth rows also represent pressure p' and minus buoyancy b, respectively. White denotes negative values; black, positive ones. For completeness, we mention the parameter values used here, although the structures are generic: buoyancy frequency $N = 1.0 \times 10^{-3}$, tidal frequency $\omega = 1.405 \times 10^{-4}$ (M_2), Coriolis parameter $f = 1.0 \times 10^{-4}$, all in rad/s, and water depth $H = 4,000$ m; distance along the horizontal and vertical axes is in kilometers.

velocity u, obtained from (7.45) and (7.46), at three different moments. We include various numbers of modes.[8] We observe a transition from the first mode alone (upper panels) to an increasingly fine structure as more higher modes come into play, from which a diagonal beam-like structure gradually emerges (toward the lowest panels). While each mode individually travels to the right at its own phase speed, their superposition remarkably produces this structure, which *stays in place* as time progresses.

---

[8] In the example of Figure 7.10, we assign equal amplitudes to the different modes. A case of decreasing amplitudes for higher modes is discussed in Section 7.6.

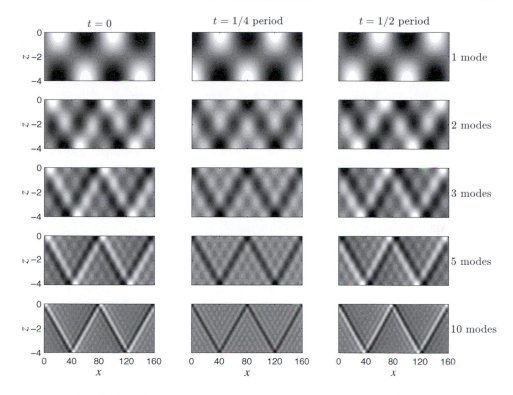

Figure 7.10 Spatial structure of superpositions of modes, for rightward propagating waves, at three different moments (from left to right). In each panel, the horizontal velocity component u is plotted, with negative values in white, and positive values in black. Parameters are as in Figure 7.9. In these superpositions, the amplitudes of the modes are chosen equal.

To understand the principle behind this orderly pattern, we reconsider the nature of the modal solution. With (7.41), we have a horizontally propagating wave whose vertical structure is described by a sinusoidal $W(z)$. We can interpret W_n as a standing wave in the vertical, i.e., a combination of up- and downward propagating waves; m_n serves as the vertical wavenumber. Now, from (7.48), we find that the ratio of the vertical and horizontal wavenumbers, m_n and k_n, is given by

$$\frac{m_n}{k_n} = \pm\left(\frac{N^2 - \omega^2}{\omega^2 - f^2}\right)^{1/2}, \tag{7.60}$$

which is *independent of mode number n*. In other words, one and the same angle in the xz-plane pervades all modes. So it is not surprising to find a well-defined pattern of diagonals in Figure 7.10. More specifically, (7.60) represents the tangent of the angle that the wave vector $\vec{k} = (k, m)$ makes with the horizontal, with the wave vector pointing in the direction of phase propagation. From the sequence of snapshots in the lowest panels of Figure 7.10, we notice that the lines of equal phase, here indicated by positive (black) and negative (white) currents, shift across the diagonals in time, which means that phase propagation is perpendicular to the diagonals themselves. Meanwhile, the levels of high energy (i.e., strong currents) stay put within the diagonal beams, so energy must propagate *along* these diagonals. We come back to these properties in Section 7.5, where they are derived in a more direct way.

In conclusion, we emphasize that the case of constant N is simple and instructive, but in some ways atypical. Two key properties – independence of the modal structure on wave frequency, and commensurability of wavenumbers, which lend much simplicity to this case – do not in general hold for vertically varying $N(z)$. We elaborate on these points in the following section.

7.4.2 Three-Layer System

We return to the stratification sketched in Figure 7.5. We assume that the wave frequency lies in the interval

$$|f| < \omega < N_c. \tag{7.61}$$

We first solve (7.42) for each layer separately, and then connect the solutions.

In the deep layer, solutions of (7.42) are sinusoidal because of (7.61),

$$W_l = A \sin q_l(z + H), \qquad \text{for } -H < z < -d - \varepsilon, \tag{7.62}$$

with an arbitrary constant A and

$$q_l = k \left(\frac{N_c^2 - \omega^2}{\omega^2 - f^2} \right)^{1/2}.$$

By leaving out the cosine term from (7.62), the boundary condition at the bottom, in (7.43), is automatically satisfied.

In the upper mixed layer, where $N = 0$, neither of the conditions (7.49) is satisfied, so the solution (7.42) is exponential and can be written as

$$W_u = B \sinh q_u z, \qquad \text{for } -d < z < 0, \tag{7.63}$$

where B is an arbitrary constant and

$$q_u = k \left(\frac{\omega^2}{\omega^2 - f^2} \right)^{1/2}.$$

Notice that (7.63) already satisfies the boundary condition at the surface, in (7.43).

We include the effect of the thermocline by taking the vertical integral of (7.42) across the thermocline, where $N^2 = g'/\varepsilon$:

$$\int_{-d-\varepsilon}^{-d} dz \left\{ W'' + k^2 \frac{N^2(z) - \omega^2}{\omega^2 - f^2} W \right\}$$

$$= W'(-d) - W'(-d - \varepsilon) + k^2 \frac{g'/\varepsilon - \omega^2}{\omega^2 - f^2} \int_{-d-\varepsilon}^{-d} dz\, W = 0. \qquad (7.64)$$

To simplify the analysis, we take the limit $\varepsilon \to 0$, which reduces the thermocline to an interface. For small ε, the integral in the last term of (7.64) can be approximated by $\varepsilon W(-d)$. Hence in the limit $\varepsilon \to 0$, (7.64) becomes

$$W_u'(-d) - W_l'(-d) + \frac{g'k^2}{\omega^2 - f^2} W(-d) = 0. \qquad (7.65)$$

Moreover, at the interface, we require continuity of W,

$$W_u(-d) = W_l(-d). \qquad (7.66)$$

Thus, for the evaluation of the last term in (7.65), it is immaterial which of the two, W_u or W_l, is taken; we choose the former. Without a thermocline ($g' = 0$), the derivative W' would be continuous, according to (7.65). The presence of an interfacial thermocline creates a discontinuity in W', and hence, by (7.46), a jump in the horizontal currents u and v across the interface. Thus, the thermocline is accompanied by a strong vertical shear in horizontal currents.

We collect (7.65) and (7.66) in a matrix

$$\begin{pmatrix} -q_l \cos q_l(H - d) & q_u \cosh q_u d - \frac{g' q_u^2}{\omega^2} \sinh q_u d \\ \sin q_l(H - d) & \sinh q_u d \end{pmatrix} \begin{pmatrix} A \\ B \end{pmatrix} = \begin{pmatrix} 0 \\ 0 \end{pmatrix}. \qquad (7.67)$$

The dispersion relation follows from the requirement that the determinant be zero:

$$q_l \cot q_l(H - d) + q_u \coth q_u d - \frac{g' q_u^2}{\omega^2} = 0. \qquad (7.68)$$

With the definitions of q_u and q_l, this forms a dispersion relation that connects ω and k. However, the equation is transcendental (i.e., cannot be solved by analytical means), so we have to resort to numerical methods to find the wavenumbers k for given tidal frequency ω. For example, we can simply calculate the left-hand side of (7.68) for a range of finely gridded k and pick out the zeros.[9] This way, we obtain wavenumbers $k_1, k_2, k_3 \cdots$, which in general turn out to be *incommensurable*.

Using the second equation in (7.67), we can express B in terms of A:

$$B = -\frac{\sin q_l(H - d)}{\sinh q_u d} A.$$

---

[9] For the numerical calculation it is better to first multiply all terms in (7.68) with $\sin q_l(H - d)$, lest problems with infinities occur.

Finally, assembling the solutions of the layers, (7.62) and (7.63), the vertical modal structure can be written

$$W(z) = A \begin{cases} -\dfrac{\sin q_l (H-d)}{\sinh q_u d} \sinh q_u z & -d < z < 0 \\ \sin q_l (z+H) & -H < z < -d. \end{cases} \tag{7.69}$$

For each wavenumber k_n, we have a corresponding q_u and q_l, amplitude A_n and vertical structure W_n. Notice that the vertical modes W_n now depend on the wave frequency ω, via q_u and q_l (in contrast to the case of constant N, in the previous section). An example of the first five vertical modes is shown in Figure 7.11a, and a superposition of 25 modes in Figure 7.11b.

In the deep layer internal tidal beams are clearly visible, but where they impinge on the thermocline, strong currents appear in the mixed layer. The beam, meanwhile, becomes less intense. The thermocline forms a strong inhomogeneity in the

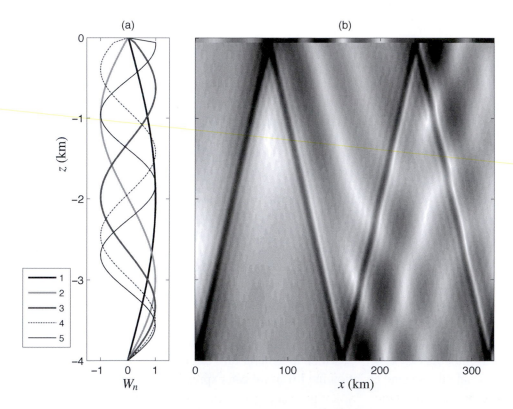

Figure 7.11 Internal tides in the three-layer system of Figure 7.5, in the limit $\varepsilon \to 0$. In (a), the first five modes of W_n are shown. In (b), a superposition of 25 modes, representing the amplitude of u. White denotes zero; black, maximum values. Parameters are $g' = 0.005$ m/s^2, mixed-layer depth $d = 60$ m; buoyancy frequency of deep layer $N_c = 2 \times 10^{-3}$, $f = 1 \times 10^{-4}$ and $\omega = 1.405 \times 10^{-4}$ (all in rad/s). Modal coefficients are $A_n = 1/n$.

stratification, and causes internal reflections of an impinging beam. Further to the right, we see broad, rather weak beams radiating downward from the thermocline. As this is a linear solution, there is no interaction among modes. It is just the superposition of modes, with their incommensurable wavenumbers, that creates this complicated pattern. In particular, there is no longer a horizontal periodicity.

Exercises

7.4.1 Consider a system consisting of two layers of equal thickness, each with constant N: N_1 (N_2) in the upper (lower) layer. Assume that $|f| < \omega < N_{1,2}$.

   a) Derive the modal structure and dispersion relation.
   b) Take the parameter values: $N_1 = 2.5 \times 10^{-3}$, $N_2 = 1 \times 10^{-3}$, $f = 1 \times 10^{-4}$ and $\omega = 1.4 \times 10^{-4}$, all in rad/s, and water depth $H = 4,000$ m. Solve the dispersion relation numerically for the first ten modes.
   c) Plot a superposition of modal solutions and interpret the behavior of the beams.

7.4.2 The same as in the previous exercise, but now with an unstratified lower layer: $N_2 = 0$ rad/s.

7.5 Characteristics

Our starting point is again (7.40), which governs the spatial structure of internal waves:

$$(f^2 - \omega^2)\frac{\partial^2 \hat{w}}{\partial z^2} + (N^2 - \omega^2)\frac{\partial^2 \hat{w}}{\partial x^2} = 0. \tag{7.70}$$

Throughout this section, we assume N to be constant. The difference from the previous section is that we now exploit the similarity in the appearances of x and z in (7.70).

7.5.1 Basic Properties of Plane Waves

The dispersion relation, which provides the connection between wave frequency and wavenumbers, can be found by substituting a plane wave solution $\hat{w} \sim \exp i(kx + mz)$ into (7.70):

$$\omega^2 = \frac{N^2 k^2 + f^2 m^2}{k^2 + m^2}. \tag{7.71}$$

This is in agreement with (7.55), but there is now no reference to vertical modes. We can simplify (7.71) by writing the wave vector in polar coordinates,

$$\vec{k} = (k, m) = \kappa(\cos\theta, \sin\theta); \quad \kappa = (k^2 + m^2)^{1/2}, \tag{7.72}$$

where κ is the length of the wave vector and θ the angle of the wave vector to the horizontal. With this, (7.71) becomes

$$\boxed{\omega^2 = N^2 \cos^2\theta + f^2 \sin^2\theta.} \tag{7.73}$$

This relation lies at the heart of internal-wave theory. It again contains the two key parameters N and f, which represent the two restoring forces at work in internal waves: buoyancy and the Coriolis force. Importantly, (7.73) shows that the wave frequency depends only on the *direction* θ of the wave vector, and not on its length κ. Conversely, we can state that for a given wave (tidal) frequency, the angle of the wave vector is fixed.

From the fact that there is no dependence on κ ($\partial\omega/\partial\kappa = 0$), we can derive another key property of internal waves. Applying the chain rule for differentiation to $\omega(k(\kappa,\theta), m(\kappa,\theta))$, we can express the derivative to κ also as

$$\frac{\partial\omega}{\partial\kappa} = \frac{\partial\omega}{\partial k}\frac{\partial k}{\partial\kappa} + \frac{\partial\omega}{\partial m}\frac{\partial m}{\partial\kappa} = \frac{\vec{c}_g \cdot \vec{k}}{\kappa}, \tag{7.74}$$

where the inner product features the group velocity vector

$$\vec{c}_g = \left(\frac{\partial\omega}{\partial k}, \frac{\partial\omega}{\partial m}\right). \tag{7.75}$$

Since the wave frequency is independent of κ, it follows from (7.74) that $\vec{c}_g \cdot \vec{k} = 0$, hence

$$\boxed{\vec{c}_g \perp \vec{k}.} \tag{7.76}$$

Another way to arrive at this property is by calculating the components of the group velocity vector explicitly from (7.71). Cast in polar coordinates, the result is

$$\vec{c}_g = \frac{(N^2 - f^2)\cos\theta\sin\theta}{\kappa\omega}(\sin\theta, -\cos\theta). \tag{7.77}$$

This not only confirms (7.76), but provides some extra information. The sign of the horizontal component of the group velocity vector is determined by $(N^2 - f^2)\cos\theta$ (the remaining factor being always positive), implying that it has the same sign as k if $N > |f|$, and the opposite sign if $N < |f|$. The sign of the vertical component of the group velocity vector is determined by $(f^2 - N^2)\sin\theta$, which has the same sign as m if $N < |f|$, and the opposite sign if $N > |f|$. We can summarize this by stating that $\vec{c}_g$ and $\vec{k}$ are vertically opposed if $N > |f|$, and horizontally opposed if $N < |f|$. The former case prevails in the ocean and is illustrated schematically in Figure 7.12.

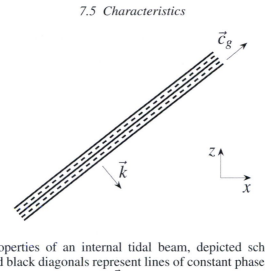

Figure 7.12 Properties of an internal tidal beam, depicted schematically. The dashed and solid black diagonals represent lines of constant phase; they propagate in the direction of the wave vector $\vec{k}$, which in this example points right- and downward. The energy propagates in the direction of the group velocity vector $\vec{c}_g$, which is along the beam and perpendicular to $\vec{k}$. In this example, we assume $N > |f|$; the vectors $\vec{k}$ and $\vec{c}_g$ then point in the same horizontal direction, but are vertically opposed.

Finally, writing the left-hand side of (7.73) as $\omega^2(\cos^2\theta + \sin^2\theta)$, we readily obtain an alternative form of the dispersion relation,

$$\cot^2\theta = \frac{\omega^2 - f^2}{N^2 - \omega^2}. \tag{7.78}$$

The right-hand side must be positive, which leads us back to the two possible regimes of the wave frequency stated in (7.49). Moreover, it follows from (7.77) that $\cot\theta$ represents the steepness of the group velocity vector. Hence, for large N (as in the thermocline), it becomes more horizontal; for weaker N (in the deeper layers of the ocean), it becomes steeper.

7.5.2 *General Solution*

Equation (7.70) can also be written as

$$\frac{\partial^2 \hat{w}}{\partial x^2} - \frac{\omega^2 - f^2}{N^2 - \omega^2} \frac{\partial^2 \hat{w}}{\partial z^2} = 0. \tag{7.79}$$

We assume that the wave frequency falls into one of the two regimes of (7.49), and define

$$\alpha = \left(\frac{\omega^2 - f^2}{N^2 - \omega^2}\right)^{1/2}. \tag{7.80}$$

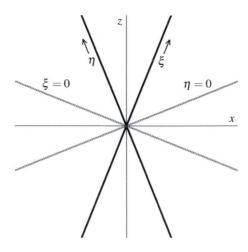

Figure 7.13 Characteristic coordinates ξ and η, defined in (7.82). Solid black lines represent the ξ and η axes, with the positive direction indicated by the arrows. Gray dashed lines are examples of ξ=const and η=const; energy propagates along those lines.

Equation (7.79) is known as the *wave equation* and its general solution is given by[10]

$$\boxed{\hat{w} = F(z+\alpha x) + G(z-\alpha x),} \tag{7.81}$$

for arbitrary functions F and G. The arguments feature the so-called characteristic coordinates

$$\xi = z + \alpha x, \qquad \eta = z - \alpha x. \tag{7.82}$$

Along lines of constant ξ, F takes a constant value, and similarly for η and G. It follows from (7.77), (7.78) and (7.80) that the group velocity vector is aligned to them (Figure 7.13).

To show an example of the solution (7.81), we take[11]

$$F(\xi) = \exp(-\xi^2)\exp ik\xi, \tag{7.83}$$

and the same functional dependence for $G(\eta)$. Together with the temporal factor $\exp(-i\omega t)$, the real part becomes

$$w = \overbrace{\exp(-\xi^2)\cos(k\xi - \omega t)}^{F} + \overbrace{\exp(-\eta^2)\cos(k\eta - \omega t)}^{G}, \tag{7.84}$$

and is illustrated in Figure 7.14, as a snapshot at $t = 0$. Notice that positive k here means upward phase propagation, corresponding to the direction of increasing ξ

[10] This can be demonstrated by transforming (7.79) to the coordinates (7.82), whence $\partial^2\hat{w}/\partial\xi\partial\eta = 0$.
[11] As our sole purpose here is to illustrate the overall structures, we casually leave aside coefficients that would need to be included to make the expressions dimensionally correct.

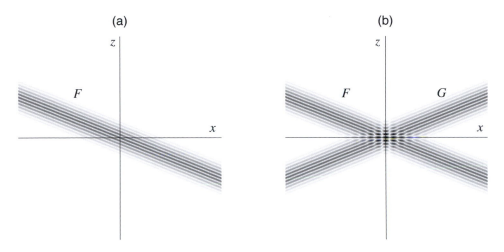

Figure 7.14 Example of solution (7.84), showing $|w|$ at $t = 0$; parameter values are $k = 20$ and $\alpha = 0.4$. In (a) F alone; in (b) superposition of F and G.

and η in Figure 7.13. The Gaussian factor gives the beams their confined width, while the cosine describes the phase propagation within the beams, in a direction perpendicular to the beams (cf. the sketch in Figure 7.12).

The solution in terms of characteristic coordinates, given by (7.81), provides a very direct and simple way to create beam-like structures. Characteristic coordinates are so suitable precisely because they are aligned to the beams. In contrast, if we cast the problem in terms of x and z, as in Section 7.4, we have to add quite a few modes to obtain an internal tidal beam (see Figure 7.10). Notice that we have so far not considered any boundary in this section; the beams are without end. We now consider what happens when they encounter a boundary, for instance a sloping bottom.

7.5.3 Reflection from a Slope

We introduce a linearly sloping bottom, defined by

$$z = \gamma x,$$

with constant γ; throughout this section we take $\gamma > 0$. As a boundary condition, we impose zero normal flow at the slope:

$$\hat{w} = \gamma \hat{u} \qquad \text{at } z = \gamma x. \tag{7.85}$$

We already have $\hat{w}$ from (7.81) and can obtain $\hat{u}$ from the continuity equation (7.26):

$$\hat{u} = -\alpha^{-1} F(z + \alpha x) + \alpha^{-1} G(z - \alpha x). \tag{7.86}$$

Hence, by applying (7.85) to $\hat{u}$ and $\hat{w}$, we obtain, after rearranging terms,

$$G([\gamma - \alpha]x) = \frac{\gamma + \alpha}{\gamma - \alpha} F([\gamma + \alpha]x).$$

Since this must hold for all x, we can express $G(\eta)$ in terms of F as

$$G(\eta) = \lambda F(\lambda \eta) \qquad \text{with } \lambda = \frac{\gamma + \alpha}{\gamma - \alpha}. \tag{7.87}$$

At this point, we already notice that λ becomes infinite when $\gamma = \alpha$; we come back to the underlying physical reason in a moment. At any rate, with (7.87) we have completed the solution (7.81) for reflecting internal waves from a slope:

$$\hat{w} = F(z + \alpha x) + \lambda F(\lambda(z - \alpha x)), \tag{7.88}$$

where F is still an arbitrary function.

In Figure 7.15a we show an example in which the slope is less steep than the beams (*subcritical* reflection, $\gamma < \alpha$), while in Figure 7.15b the slope is steeper than the beams (*supercritical* reflection, $\gamma > \alpha$). In the former case, the direction of energy propagation reverses in the vertical; in the latter case, in the horizontal.

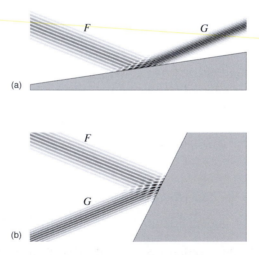

Figure 7.15 Reflection from a uniform slope, showing $|\hat{w}|$ from the solution (7.88), again with F prescribed by (7.83). In both panels, $\alpha = 0.4$ and $k = 10$. In (a), a case of subcritical reflection with slope $\gamma = 0.15$; in (b), a case of supercritical reflection with $\gamma = 2$. As k is chosen positive, phase propagation is upward for F in both cases. For G in (7.87), phase propagation is determined by λk, which is negative in (a) and positive in (b). If we assume $N > |f|$, then the group velocity vector is vertically opposed to the direction of phase propagation, so in both panels F is to be interpreted as the incident beam and G as the reflected one. If $N < |f|$, the roles are reversed.

As explained in the caption of Figure 7.15, it depends on the sign of k and on whether $|f| \lessgtr N$, which of the two beams is to be interpreted as incident and which as reflected. Regardless, we observe that the intensity of the beam changes upon reflection. Meanwhile, the (absolute) angle with the vertical stays the same, as dictated by (7.78): for a given frequency ω, the angle θ is specified. It defines the angle of $\vec{k}$ with the horizontal, which also is the angle of $\vec{c}_g$ with the vertical.

Finally, we notice that the case of critical reflection ($\gamma = \alpha$) gives rise to a beam that is aligned to the bottom slope and is of infinite intensity according to (7.87). The solution then becomes invalid, which signals that one (or more) of the underlying approximations no longer holds. In this case, an indefinite increase in amplitude invalidates the assumption of linearity but also the neglect of friction. Both come into play in any real situation of critical reflection. As in the case of the shallow-water constituents (Section 6.3), nonlinear terms give rise to overtides, i.e. tides at multiples of the basic frequency, to the extent that they fall within the window of allowable frequencies (7.49). The intense signal occurring at critical reflection also affects the local turbulent mixing. This marks bottom slopes as a favorable spot for mixing in the deep ocean.

To see how the steepness of internal tidal beams compares with the steepness of slopes in the ocean, we first examine the distribution of slopes. The procedure to calculate slopes from bathymetric data is explained in Box 7.1 and a representative example of a distribution is shown in Figure 7.16. It is clearly skewed toward mild slopes; the median value of γ is approximately 0.02. For semidiurnal internal tides (1.4×10^{-4} rad/s) at midlatitudes ($f = 1.0 \times 10^{-4}$ rad/s), we find a range of values for α, depending on the stratification. For strong stratification ($N = 1 \times 10^{-2}$ rad/s), internal tidal beams are fairly horizontal with $\alpha = 0.01$. This matches bottom slopes in the head of the distribution, but in reality their occurrence is of course rare, since the bottom is mostly located far below the thermocline (Figure 7.4), with the important exception of the shelf edge, which is the area of the transition between the continental slope and the shelf (Figure 5.3). Layers of weaker stratification ($N = 1 \times 10^{-3}$ rad/s, say) more often intersect the bottom, but now α moves to the tail of the distribution, as $\alpha = 0.1$. Thus, reflection of semidiurnal internal tides is predominantly subcritical (i.e., $\gamma < \alpha$).

While the focus in this section was on reflection from a sloping bottom, the special case of a flat bottom ($\gamma = 0$) is also of interest. In that case, solution (7.88) still applies, of course. We may add an upper surface as well, closing the fluid at the top with a rigid-lid. It can be shown that this extra condition translates into a periodicity of F. In fact, it naturally leads us back to the solution in terms of vertical modes of Section 7.4.1. After all, both deal with the same set of equations for the same configuration. The two ways of looking at the problem – vertical modes and characteristics – are then necessarily equivalent.

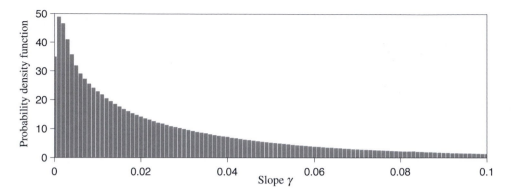

Figure 7.16 Probability density function of bottom slope γ, derived from topographic data in the zonal band between $20°$ and $40°$N in the Atlantic Ocean, using a later version of the database by Smith and Sandwell (1997), at $1/30°$ resolution.

Exercise

7.5.1 Demonstrate that the solution (7.88), applied to a flat bottom, can be written in terms of vertical modes if one adds a boundary in the form of a rigid lid at the surface.

Box 7.1 **Slopes in Ocean Bathymetry**

Here we outline the procedure to calculate the slope of tangent planes in a three-dimensional bathymetry. Let the bottom be described by $z = -H(x,y)$. It turns out to be convenient to introduce the function $\tau = z + H$, which automatically satisfies

$$\tau(x,y,z) = 0.$$

Its gradient can be written

$$\nabla\tau = \begin{pmatrix} \partial\tau/\partial x \\ \partial\tau/\partial y \\ \partial\tau/\partial z \end{pmatrix} = \begin{pmatrix} \partial H/\partial x \\ \partial H/\partial y \\ 1 \end{pmatrix},$$

which is the vector normal to the surface $\tau(x,y,z) = 0$ and hence normal to the bathymetry, since $\tau = 0$ and $z = -H$ describe the same surface.

Now, the angle of the tangent plane with the horizontal, β say, is the same as the angle between the normal and the vertical. Hence, by taking the inner product of $\nabla\tau$ with the vertical unit vector $(0,0,1)$, we obtain

$$|\nabla\tau|\cos\beta = 1.$$

The tangent of angle β is the slope γ that we seek to determine; applying Pythagorean theorem, we find the expression

$$\gamma = (|\nabla \tau|^2 - 1)^{1/2}. \tag{7.89}$$

7.5.4 Beyond Uniform Stratification

Throughout this section, we have assumed uniform N. As we have seen, this leads to internal-wave beams whose energy keeps on moving along one of the characteristic coordinates, until it meets a solid boundary. At the juncture from incident to reflected beams, the path of energy propagation changes to the other characteristic coordinate.

We can intuitively understand how this situation changes if one allows N to vary in the vertical. First, the steepness of the beam depends on N, according to (7.78). For large N (in the thermocline, say), the beam is relatively horizontal, whereas in weakly stratified layers (e.g., the deep ocean), it becomes more vertical. This means that the beam, instead of following straight lines, becomes bended or *refracted*. This aspect can be coped with by allowing the characteristic coordinates to be curved, although this obviously complicates the mathematical description.

There is, however, still another effect. In a way, the vertical variation in stratification acts as an infinite amount of soft boundaries within the fluid, which are largely transparent, but not entirely. Most of the energy of a beam generally passes through, but part of it is reflected. These *internal reflections*, which occur throughout the interior of the fluid but predominantly where stratification changes strongly, scatter the energy of an initially focused beam. Figure 7.17 sketches the idea. The occurrence of these internal reflections is the principal reason why the method of characteristics no longer offers an expedient way for the description of internal wave propagation in a nonuniform medium.

Interestingly, the method of vertical modes assimilates the phenomenon of internal reflection without any difficulty. In fact, we have already seen an example in Figure 7.11b, where the scattering of the initial beam is visible, caused by partial reflections from the thermocline.

7.6 Generation of Internal Tides

To conclude this chapter, we examine the mechanism that is responsible for the generation of internal tides. Three ingredients are needed: a bottom slope, a vertical density stratification, and a cross-slope barotropic tidal current. At the

etc.

Figure 7.17 Sketch of internal reflections of a beam entering from the upper left, in a medium where N decreases downward. The reflected beams are themselves also subject to partial reflections, so the process is endless. Note that only a few specific points of reflection have been drawn, by way of example. In reality, the process happens at any point in layers of nonuniform N.

end of Section 5.7, we have seen that such cross-slope currents will generally be present over the continental slope, for example. The ubiquitousness of the three ingredients explains why internal tides are found throughout the oceans.

7.6.1 Barotropic Forcing Term

Our starting point is again (7.37),

$$\frac{\partial^2}{\partial t^2}\nabla^2 w + f^2\frac{\partial^2 w}{\partial z^2} + N^2\nabla_h^2 w = 0. \tag{7.90}$$

To simplify matters, we assume uniformity in one of the horizontal directions, $\partial/\partial y = 0$. This allows us to introduce a stream function ψ, defined via $u = -\partial\psi/\partial z$ and $w = \partial\psi/\partial x$, for which the continuity equation (7.26) is automatically satisfied. In terms of the stream function, (7.90) becomes

$$\frac{\partial^2}{\partial t^2}\nabla^2 \psi + f^2\frac{\partial^2 \psi}{\partial z^2} + N^2\frac{\partial^2 \psi}{\partial x^2} = 0, \tag{7.91}$$

where now $\nabla^2 = \partial^2/\partial x^2 + \partial^2/\partial z^2$. For the moment, we define the bottom by

$$z = -H(x).$$

In previous sections, we have used (7.90) exclusively as a tool to describe internal tides. However, its derivation imposes no such restriction and the equation may be used to describe barotropic tides as well. They will be prescribed, but without considering the details of their propagation as a wave. In fact, we exclude surface movements altogether by retaining a rigid-lid at the surface. The barotropic tide is thus stripped of most of its dynamics except its flow, which is the essential element in the generation of internal tides. Specifically, we conceive the horizontal barotropic tidal current U as a simple oscillatory flow, constrained between the rigid boundaries below and on top:

$$U(t,x) = \frac{Q_0}{H(x)} \exp(-i\omega t), \qquad (7.92)$$

with constant flux amplitude Q_0. The strength of U is inversely proportional to the local water depth $H(x)$; the shallower, the stronger the flow. In terms of the stream function, we can express the barotropic tidal flow (7.92) as

$$\Psi_f = -\frac{zQ_0}{H(x)} \exp(-i\omega t). \qquad (7.93)$$

Importantly, (7.93) implies that there is also a *vertical* barotropic tidal current over the bottom slope, $W = \partial \Psi_f / \partial x$. This oscillatory vertical current periodically lifts up and pulls down the isopycnals. It is through this movement that internal tides are generated; they will be represented by Ψ. Substituting

$$\psi = \Psi_f + \Psi$$

into (7.91) gives

$$\frac{\partial^2}{\partial t^2}\nabla^2\Psi + f^2\frac{\partial^2\Psi}{\partial z^2} + N^2\frac{\partial^2\Psi}{\partial x^2} = zQ_0(N^2 - \omega^2)\frac{d^2}{dx^2}\left(\frac{1}{H}\right)\exp(-i\omega t). \qquad (7.94)$$

For Ψ, we impose the following boundary conditions:

$$\Psi = 0 \quad \text{at } z = -H(x), 0. \qquad (7.95)$$

The right-hand side of (7.94) contains the barotropic term, but it is actually in combination with the boundary conditions (7.95) that the term takes its role as a body forcing for internal tides. The point is that Ψ was hitherto unspecified; only through (7.95) is its baroclinic character imposed,[12] for we now have

$$\int_{-H}^{0} \Psi_z = 0,$$

i.e., the vertically integrated horizontal current is zero, a distinctive characteristic of internal waves (cf. (7.47)).

---

[12] In the present nonhydrostatic setting there is a subtle point, for Ψ can be shown to contain also a nonhydrostatic contribution to the barotropic field; see Garrett and Gerkema (2007) for the details.

We simplify the problem stated in (7.94) and (7.95) by assuming that the bathymetric form (e.g., a bank or ridge) can be regarded as a perturbation, in the sense that its height is small compared to the water depth. The bottom is now described by

$$z = -H(x) = -H_0 + h(x), \qquad \text{with } |h| \ll H_0, \tag{7.96}$$

where H_0 is the unperturbed water depth. Using the binomial series (A.9), we make the approximation

$$\frac{1}{H(x)} = \frac{1}{H_0(1 - h/H_0)} = \frac{1}{H_0}\left(1 + \frac{h}{H_0} + \cdots\right).$$

Substitution in (7.94) gives

$$\frac{\partial^2}{\partial t^2}\nabla^2\Psi + f^2\frac{\partial^2\Psi}{\partial z^2} + N^2\frac{\partial^2\Psi}{\partial x^2} = \frac{zQ_0}{H_0^2}(N^2 - \omega^2)\frac{d^2h}{dx^2}\exp(-i\omega t). \tag{7.97}$$

The boundary condition for the bottom is now applied at $z = -H_0$, so (7.95) becomes

$$\Psi = 0 \quad \text{at } z = -H_0, 0. \tag{7.98}$$

This greatly simplifies the problem, for we now have effectively a flat bottom, which means that we can solve the problem in terms of vertical modes, as in Section 7.4.

7.6.2 General Solution

For specific choices of stratification $N(z)$ and bathymetry $h(x)$, (7.97) and (7.98) can be solved. However, we first proceed without specifying them, to derive a general expression for the solution.

Following the results of Section 7.4, we anticipate that Ψ can be expressed as a sum of vertical modes:

$$\Psi = \sum_n a_n(x)\Phi_n(z)\exp(-i\omega t), \tag{7.99}$$

with complex a_n. The vertical modes Φ_n are solutions of (7.42),

$$\Phi_n'' + k_n^2\frac{N^2(z) - \omega^2}{\omega^2 - f^2}\Phi_n = 0. \tag{7.100}$$

We substitute (7.99) in (7.97); this gives

$$\sum_n \frac{N^2 - \omega^2}{\omega^2 - f^2}\Phi_n\frac{d^2a_n}{dx^2} - \sum_n \Phi_n'' a_n = \frac{zQ_0}{H_0^2}\frac{N^2 - \omega^2}{\omega^2 - f^2}\frac{d^2h}{dx^2}. \tag{7.101}$$

We multiply this equation with vertical mode Φ_j and use (7.100), applied to Φ_j, to simplify the first term on the left-hand side and the forcing term,

$$\sum_n \Phi_n \Phi_j'' \frac{d^2 a_n}{dx^2} + k_j^2 \sum_n \Phi_n'' \Phi_j a_n = \frac{z Q_0}{H_0^2} \Phi_j'' \frac{d^2 h}{dx^2}. \tag{7.102}$$

We now take the integral over the vertical and use the orthogonality properties (7.44), which imply that the only contribution comes from $n = j$, since otherwise the terms on the left-hand side vanish identically. Hence

$$\frac{d^2 a_n}{dx^2} + k_n^2 a_n = \frac{d_n Q_0}{H_0^2} \frac{d^2 h}{dx^2}, \tag{7.103}$$

where the coefficient d_n is defined by

$$d_n = \frac{\int_{-H_0}^0 dz\, z\, \Phi_n''}{\int_{-H_0}^0 dz\, \Phi_n'' \Phi_n}. \tag{7.104}$$

The problem has thus been reduced to solving (7.100) and (7.103). In Section 7.4 we already discussed examples of profiles $N(z)$ for which (7.100) can be solved analytically. The other equation, (7.103), is also easy to solve; its general solution reads

$$a_n = C_{1,n} \exp(i k_n x) + C_{2,n} \exp(-i k_n x) + \frac{d_n Q_0}{H_0^2} \int_0^x d\bar{x}\, \cos k_n (x - \bar{x}) \frac{dh}{d\bar{x}}, \tag{7.105}$$

in which the first two terms (with arbitrary complex coefficients $C_{1,n}$ and $C_{2,n}$) describe free left- and rightward propagating waves. The third term is a particular solution of (7.103), as can be verified by substitution. In the integral, h features as a function of $\bar{x}$.

However, (7.105) is not yet by itself an appropriate solution to our problem. The general solution contains internal tides traveling away from the forcing region as well as ones traveling towards it. We need to exclude the latter, since our interest lies in internal tides that are generated in the forcing region. This is done by imposing *radiation conditions*. For $k_n > 0$, they read[13]

$$a_n \sim \begin{cases} \exp(+i k_n x) & \text{as } x \to +\infty \\ \exp(-i k_n x) & \text{as } x \to -\infty. \end{cases} \tag{7.106}$$

These requirements allow us to determine the constants $C_{1,n}$ and $C_{2,n}$. First, in the integral in (7.105), we write the cosine in terms of complex exponentials, as in (6.75). Hence a_n becomes

---

[13] NB: The radiation conditions actually concern the direction of *energy* propagation. We implicitly assume here that $|f| < N$, in which case phases and energy have the same horizontal direction of propagation, see (7.56) and (7.57); we may then pose the conditions in terms of phase propagation.

$$a_n = \left[C_{1,n} + \frac{d_n Q_0}{2H_0^2} \int_0^x d\bar{x} \exp(-ik_n\bar{x}) \frac{dh}{d\bar{x}} \right] \exp(ik_nx)$$

$$+ \left[C_{2,n} + \frac{d_n Q_0}{2H_0^2} \int_0^x d\bar{x} \exp(ik_n\bar{x}) \frac{dh}{d\bar{x}} \right] \exp(-ik_nx).$$

To satisfy the radiation conditions (7.106), we must choose

$$C_{1,n} = \frac{d_n Q_0}{2H_0^2} \int_{-\infty}^0 d\bar{x} \exp(-ik_n\bar{x}) \frac{dh}{d\bar{x}}, \quad C_{2,n} = -\frac{d_n Q_0}{2H_0^2} \int_0^\infty d\bar{x} \exp(ik_n\bar{x}) \frac{dh}{d\bar{x}}.$$

Hence

$$a_n = \frac{d_n Q_0}{2H_0^2} \left[\tilde{A}_n(x) \exp(ik_nx) - \tilde{B}_n(x) \exp(-ik_nx) \right], \qquad (7.107)$$

with

$$\tilde{A}_n(x) = \int_{-\infty}^x d\bar{x} \exp(-ik_n\bar{x}) \frac{dh}{d\bar{x}}, \qquad \tilde{B}_n(x) = \int_x^\infty d\bar{x} \exp(ik_n\bar{x}) \frac{dh}{d\bar{x}}. \quad (7.108)$$

With this, a closed solution satisfying the radiation conditions has been obtained for general small-amplitude bathymetry. For specific choices of h, the integrals $\tilde{A}$ and $\tilde{B}$ can be evaluated numerically, and in some special cases analytically.

7.6.3 Example

As an example, we consider a symmetric seamount, described by a continuously differentiable function:

$$h(x) = \begin{cases} 0 & \text{for } |x| > L \\ h_0(x^2/L^2 - 1)^2 & \text{for } |x| < L, \end{cases} \qquad (7.109)$$

with amplitude h_0 and width $2L$ (Figure 7.18).

For the stratification, we take the simplest choice: a constant N. This leads us back to Section 7.4.1, with k_n given by (7.53), here taken positive. The vertical modes Ψ_n are given by (7.58), with Ψ_n replacing W_n. The choice of A_n is immaterial (as long as it is not zero), because the factor will be compensated in a_n, via

Figure 7.18 A seamount, defined by (7.109).

d_n in (7.104); amplitudes are now imposed by the forcing. We can therefore simply take $A_n = 1$ for all n in (7.58). Evaluation of (7.104) gives

$$d_n = -\frac{2H_0}{n\pi}(-1)^n. \qquad (7.110)$$

For a specific choice of parameters H_0, h_0, L, N, ω and f, we can numerically evaluate (7.108). Notice that they need only be calculated within the range $x = (-L, L)$, as they are constant outside of this interval (zero at one side). We can then calculate a_n from (7.107) and Ψ from (7.99), from which u follows as

$$u = -\sum_n a_n(x)\Phi'_n(z)\exp(-i\omega t). \qquad (7.111)$$

We plot the solution in five snapshots, covering half a tidal period (Figure 7.19). Two beams emanate from the center, where the seamount is located: one to the

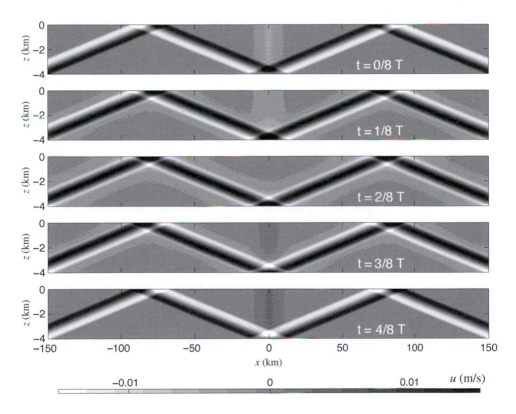

Figure 7.19 Example of internal tide generation over the seamount shown in Figure 7.18, for a superposition of 25 modes. Here $L = 15$ km and $h_0 = 500$ m. Other parameters are: $Q_0 = 100$ m^2/s, constant $N = 2 \times 10^{-3}$ rad/s, $f = 1 \times 10^{-4}$ rad/s and $\omega = 1.405 \times 10^{-4}$ rad/s (M$_2$). Five snapshots are shown, at intervals of $T/8$ (T the tidal period).

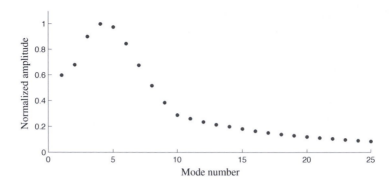

Figure 7.20 Contribution of individual modes to (7.111). Parameters are as in Figure 7.19.

left, the other to the right. The direction of phase propagation can be deduced by comparing subsequent panels. Since $N > |f|$, the direction of the group velocity is horizontally the same as the phase speed, but vertically opposed (as discussed in Section 7.5.1). This correctly implies that energy propagates away from the source, to the left for negative x and to the right for positive x. The width of the beam is set by the horizontal size of the seamount (its height h_0 here merely serves as a multiplication factor in the strength of the beam, as we assume the seamount to be infinitesimal).

In this example, 25 modes are included. We plot their relative importance in Figure 7.20. High mode numbers only give a minor contribution, while the largest contribution comes from modes 4 and 5. This peak shifts as one changes the length of the slope L: for longer slopes, the lower modes, and ultimately mode 1, become the most important, while for shorter slopes, the peak shifts to higher modes. The optimal generation takes place for modes whose wavelength broadly matches the length of the seamount ($2L$).

7.6.4 Numerical Model Results

The example shown in the previous section is instructive but of course academic, in view of all the simplifications made. In reality, the important regions of internal tide generation are often characterized by high and steep bathymetry and by complex profiles of varying N. In such cases, one has to resort to numerical modeling.

We show here an example of a model result, obtained by solving numerically the linear equations (7.28) to (7.32), but now with a barotropic forcing term $W = \partial \Psi_f / \partial x$ on the right-hand side of (7.31):

$$\frac{\partial b}{\partial t} + N^2 w = -N^2 W. \tag{7.112}$$

The hydrostatic approximation is made (i.e., neglecting $\partial w/\partial t$ in (7.30)) and uniformity in the y direction is assumed ($\partial/\partial y = 0$). The bathymetry $h(x)$ and stratification $N(z)$ are chosen in accordance with empirical profiles. In the present example, those profiles are taken from the Great Meteor Seamount, which lies in the western part of the Canary Basin, halfway between the Canary Islands and the Mid-Atlantic Ridge. It is a guyot, named after the research vessel "Meteor" with which it was discovered in 1938. The model is forced at the M_2 frequency.

The amplitude of u is shown in Figure 7.21a. A number of elements treated in this chapter can be recognized. First of all, we observe beam-like structures, emanating from the top of the seamount. As they come in deeper water, they become steeper. This refraction occurs because the stratification in the deep ocean is weaker, as discussed in Section 7.5.4. The pattern as a whole is complex since the beams seem to be everywhere. This is quite unlike the very clean beams depicted in Figure 7.19. However, it is not difficult to find the key factor responsible for this

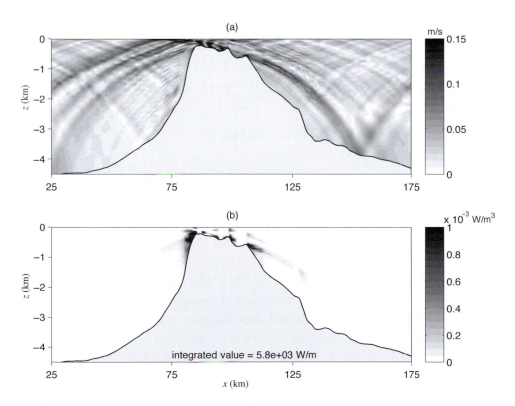

Figure 7.21 Numerical model results for internal-tide generation over the Great Meteor Seamount. In (a), the amplitude of current velocity u. In (b), the conversion rate. Its spatially integrated value is also stated. Based on Gerkema and Van Haren (2007).

messiness. After all, the mere presence of an interfacial thermocline in Figure 7.11b already produced a spatial spreading of beams. The strong inhomogeneity in N, especially in and around the thermocline, causes internal reflections and a scattering of beams (Section 7.5.4).

If we follow the main beam on the right, we notice that it undergoes a subcritical reflection from the seafloor, qualitatively comparable to Figure 7.15a. In short, despite the complexity in this more realistic setting, we recognize characteristic elements of internal tides that were derived from the basic theory in previous sections.

Finally, we briefly consider the energy conversion from barotropic to baroclinic tides. From the model equations used in Figure 7.21, as outlined at the beginning of this section, one can construct the following energy equation:

$$\frac{\partial E}{\partial t} + \frac{\partial}{\partial x}(up') + \frac{\partial}{\partial z}(wp') = -\rho_* b W, \qquad (7.113)$$

with energy E defined as

$$E = \tfrac{1}{2}\rho_*(u^2 + v^2 + b^2/N^2).$$

Taking a tidal average (denoted by $\langle \cdot \rangle$), the first term on the left-hand side of (7.113) vanishes if the state is periodic. The tidal average of the term on the right-hand side represents the local conversion rate C:

$$C = -\rho_* \langle bW \rangle, \qquad (7.114)$$

which is expressed in W/m^3 and depicted in Figure 7.21b. It shows where the energy is transferred from the barotropic to the baroclinic tide. High values of C occur over the top of the seamount. This has partly to do with the shallowness of the region, which gives stronger barotropic flows, and also with the locally high values of N around the thermocline (notice that N^2 appears in the forcing term in (7.112)).

The spatially integrated value of C, i.e. integrated over the xz-plane, comes in W/m. The "per meter" stands for the along-slope y direction, for which in the present setting uniformity was assumed. Integrating over the circumference of the seamount gives the estimated power associated with the conversion, 0.36 GW. This is a rough estimate, but serves as an indication of the order of magnitude. A small spot on the map, the Great Meteor Seamount offers its commensurate contribution to the global energy-flow diagram shown in Figure 1.11.

A final point of attention is the role of other tidal frequencies. In this chapter, we have only considered one tidal frequency, but in reality more constituents are present at the same time. For internal tides, the steepness of beams is set by their frequency. So, for example, a solar semidiurnal S_2 internal tidal beam propagates at a slightly steeper angle than does the lunar M_2; the beams follow different paths. Consequently, the timing and strength of the spring-neap cycle in internal tides will vary spatially.

Further Reading

The various equations of state (for density ρ, speed of sound c_s, etc.) all derive in a self-consistent way from the equation of state for the Gibbs potential (Feistel and Hagen, 1995). More documentation and scripts can be found on the website www.teos-10.org.

Vertical modal solutions for various profiles of buoyancy frequency N can be found in Roberts (1975); the book also contains an exhaustive bibliography on internal gravity waves up to the year of publication. Vlasenko et al. (2005) is entirely devoted to internal tides, with a focus on numerical modeling and non-linear evolution. An overview of observations on internal tides around the globe is presented by Morozov (2018).

In several chapters of their book, LeBlond and Mysak (1978) discuss internal wave theory, including the effects of the full Coriolis force (see Figure 5.17). In this chapter, we have used the Traditional Approximation, but the methods in Sections 7.4 and 7.5 can be generalized in a straightforward way to include the "non-traditional" terms (Gerkema and Shrira, 2005). These terms can become significant in weakly stratified layers.

This chapter was restricted to small-amplitude waves (linear theory). In the ocean, however, internal tides may steepen as they propagate away from their area of generation, and eventually break up into trains of high-amplitude internal solitary waves (or solitons), notably in the thermocline. An overview on the generation mechanisms can be found in Jackson et al. (2012); for their surface manifestation and observation by synthetic aperture radar imagery, see Jackson et al. (2013).

In Section 7.6 we outlined the principle of internal tidal generation by means of a simple example. Garrett and Kunze (2007) discuss the various parameter regimes involved in the generation process. Numerical models are an indispensable tool for studying the generation of internal tides in the presence of complex bathymetry and stratification; Carter et al. (2012) present an overview of the various kinds of numerical models that have been developed.

Maps of barotropic tidal dissipation are presented by Egbert and Ray (2003b), based on satellite altimeter data. In the deep ocean, the dissipation is largely due to a conversion to internal tides. However, the scales of internal tides are too small to capture in global ocean models. Hence, methods have been developed to parameterize the conversion, i.e., find approximate representations without explicitly modeling them. Green and Nycander (2013) give a review of different methods. Parameterizations have been included in global ocean circulation models; the spatial distribution of hot spots of high vertical diffusivity and mixing turns out to be an important factor, as shown, e.g., by Simmons et al. (2004). A general survey on deep ocean mixing, including the role of internal tides, can be found in Garrett and St. Laurent (2002).

Appendix A

Mathematical Formulae

For convenience, we list the trigonometric identities that are used at various places in this book:

$$\sin(\alpha \pm \beta) = \sin\alpha\cos\beta \pm \cos\alpha\sin\beta \tag{A.1}$$

$$\cos(\alpha \pm \beta) = \cos\alpha\cos\beta \mp \sin\alpha\sin\beta \tag{A.2}$$

$$2\cos\alpha\cos\beta = \cos(\alpha - \beta) + \cos(\alpha + \beta) \tag{A.3}$$

$$2\sin\alpha\sin\beta = \cos(\alpha - \beta) - \cos(\alpha + \beta) \tag{A.4}$$

$$2\sin\alpha\cos\beta = \sin(\alpha - \beta) + \sin(\alpha + \beta). \tag{A.5}$$

Special cases are

$$\cos 2\alpha = 2\cos^2\alpha - 1 \tag{A.6}$$

$$= 1 - 2\sin^2\alpha, \tag{A.7}$$

$$\sin 2\alpha = 2\sin\alpha\cos\alpha. \tag{A.8}$$

The following Taylor series are also used:

$$(1+x)^p = 1 + px + \frac{p(p-1)}{2!}x^2 + \frac{p(p-1)(p-2)}{3!}x^3 + \cdots \tag{A.9}$$

$$\sin x = x - \frac{x^3}{3!} + \frac{x^5}{5!} + \cdots \tag{A.10}$$

$$\cos x = 1 - \frac{x^2}{2!} + \frac{x^4}{4!} + \cdots . \tag{A.11}$$

Equation (A.9) is known as the *binomial series*. In all three cases, the first few terms already provide a good approximation if $|x| \ll 1$.

For the derivation of the depth-averaged momentum and continuity equations (see Appendix B), we need Leibniz' rule of integration

$$\frac{\partial}{\partial x}\int_{a(x)}^{b(x)} f(x,z)\,dz = \int_{a(x)}^{b(x)} \frac{\partial f}{\partial x}(x,z)\,dz + f(x,b(x))\frac{db}{dx} - f(x,a(x))\frac{da}{dx}. \tag{A.12}$$

Appendix B

Depth-Averaged Shallow-Water Equations

We give an outline of the derivation of the depth-averaged shallow-water equations (6.13), (6.14), and (6.15). They are expressed in terms of the depth-averaged horizontal currents defined in (6.12). The depth-averaging is across the vertical defined by the bottom at $z = -H(x, y)$ and the surface $z = \zeta(t, x, y)$. The boundaries are thus themselves functions of x and y and, in the case of ζ, of time. To prevent the expressions from becoming unwieldy, we denote quantities evaluated at the surface $z = \zeta$ or bottom $z = -H$ by an asterisk written as upper or lower index, respectively; so, e.g., for u:

$$u^* = u(t, x, y, \zeta), \qquad u_* = u(t, x, y, -H).$$

The boundary conditions (6.4) and (6.5) now read

$$w^* = \frac{\partial \zeta}{\partial t} + u^* \frac{\partial \zeta}{\partial x} + v^* \frac{\partial \zeta}{\partial y} \tag{B.1}$$

$$w_* = -u_* \frac{\partial H}{\partial x} - v_* \frac{\partial H}{\partial y}. \tag{B.2}$$

We start with the continuity equation (6.3):

$$\frac{\partial u}{\partial x} + \frac{\partial v}{\partial y} + \frac{\partial w}{\partial z} = 0. \tag{B.3}$$

According to Leibniz' rule for integration (A.12), we have

$$\int_{-H}^{\zeta} \frac{\partial u}{\partial x} \, dz = \frac{\partial}{\partial x} \int_{-H}^{\zeta} u \, dz - u^* \frac{\partial \zeta}{\partial x} - u_* \frac{\partial H}{\partial x}.$$

An identical expression is found for v, where we replace x with y. Vertical integration of (B.3) thus yields

$$\frac{\partial}{\partial x} \int_{-H}^{\zeta} u \, dz - u^* \frac{\partial \zeta}{\partial x} - u_* \frac{\partial H}{\partial x} + \frac{\partial}{\partial y} \int_{-H}^{\zeta} v \, dz - v^* \frac{\partial \zeta}{\partial y} - v_* \frac{\partial H}{\partial y} + w^* - w_* = 0.$$

Using the boundary conditions (B.1) and (B.2), and the definition (6.12), we obtain the depth-averaged version of the continuity equation (6.13):

$$\frac{\partial \zeta}{\partial t} + \frac{\partial}{\partial x}[(H + \zeta)\bar{u}] + \frac{\partial}{\partial y}[(H + \zeta)\bar{v}] = 0. \tag{B.4}$$

This equation is exact: the steps from (B.3) to (B.4) involve no approximations.

With regard to the horizontal momentum equations, we elaborate only on (6.10), since the procedure for (6.11) is analogous. Thus, the starting point is (6.10):

$$\frac{\partial u}{\partial t} + u\frac{\partial u}{\partial x} + v\frac{\partial u}{\partial y} + w\frac{\partial u}{\partial z} - fv = -g\frac{\partial \zeta}{\partial x} + \frac{1}{\rho}\frac{\partial F_x}{\partial z}. \tag{B.5}$$

According to Leibniz's rule for integration (A.12), we have the identity

$$\int_{-H}^{\zeta} \frac{\partial u}{\partial t}\,dz = \frac{\partial}{\partial t}\int_{-H}^{\zeta} u\,dz - u^*\frac{\partial \zeta}{\partial t}.$$

Hence, the vertically integrated version of (B.5) reads

$$\frac{\partial}{\partial t}[(H+\zeta)\bar{u}] - u^*\frac{\partial \zeta}{\partial t} + \int_{-H}^{\zeta}\left[u\frac{\partial u}{\partial x} + v\frac{\partial u}{\partial y} + w\frac{\partial u}{\partial z}\right]dz - f(H+\zeta)\bar{v}$$
$$= -g(H+\zeta)\frac{\partial \zeta}{\partial x} - \frac{\tau_x}{\rho}, \tag{B.6}$$

where we used the boundary conditions for F_x, (6.6) and (6.7). The advective terms are still to handled. The continuity equation (B.3) allows them to be written as

$$u\frac{\partial u}{\partial x} + v\frac{\partial u}{\partial y} + w\frac{\partial u}{\partial z} = \frac{\partial (u^2)}{\partial x} + \frac{\partial (uv)}{\partial y} + \frac{\partial (uw)}{\partial z}.$$

Vertical integration of the right-hand side gives, using Leibniz' rule (A.12),

$$\int_{-H}^{\zeta}\left[\frac{\partial (u^2)}{\partial x} + \frac{\partial (uv)}{\partial y} + \frac{\partial (uw)}{\partial z}\right]dz = \frac{\partial}{\partial x}\int_{-H}^{\zeta} u^2\,dz - (u^2)^*\frac{\partial \zeta}{\partial x} - (u^2)_*\frac{\partial H}{\partial x}$$
$$+ \frac{\partial}{\partial y}\int_{-H}^{\zeta} uv\,dz - (uv)^*\frac{\partial \zeta}{\partial y} - (uv)_*\frac{\partial H}{\partial y} + (uw)^* - (uw)_*$$
$$= \frac{\partial}{\partial x}\int_{-H}^{\zeta} u^2\,dz + \frac{\partial}{\partial y}\int_{-H}^{\zeta} uv\,dz + u^*\frac{\partial \zeta}{\partial t},$$

where we used the boundary conditions (B.1) and (B.2) in the last step. Substitution in (B.6) gives

$$\frac{\partial}{\partial t}[(H+\zeta)\bar{u}] + \frac{\partial}{\partial x}\int_{-H}^{\zeta} u^2\,dz + \frac{\partial}{\partial y}\int_{-H}^{\zeta} uv\,dz - f(H+\zeta)\bar{v} = -g(H+\zeta)\frac{\partial \zeta}{\partial x} - \frac{\tau_x}{\rho}. \tag{B.7}$$

The problem is now that we are left with integrals of the products u^2 and uv which cannot be expressed in terms of $\bar{u}$ and $\bar{v}$ alone. There is no perfect solution to this problem. We proceed by decomposing u and v in a depth-averaged part and the remainder (indicated by a hat):

$$u = \bar{u} + \hat{u}, \qquad v = \bar{v} + \hat{v}.$$

By definition, the vertical integrals of $\hat{u}$ and $\hat{v}$ are zero, hence

$$\int_{-H}^{\zeta} u^2\,dz = (H+\zeta)(\bar{u}^2 + \overline{\hat{u}^2})$$

$$\int_{-H}^{\zeta} uv\,dz = (H+\zeta)(\bar{u}\bar{v} + \overline{\hat{u}\hat{v}}).$$

Substitution in (B.7) gives

$$\frac{\partial}{\partial t}[(H+\zeta)\bar{u}] + \frac{\partial}{\partial x}\left[(H+\zeta)(\bar{u}^2+\overline{\hat{u}^2})\right] + \frac{\partial}{\partial y}\left[(H+\zeta)(\bar{u}\bar{v}+\overline{\hat{u}\hat{v}})\right] - f(H+\zeta)\bar{v}$$

$$= -g\,(H+\zeta)\frac{\partial\zeta}{\partial x} - \frac{\tau_x}{\rho}. \tag{B.8}$$

We simplify (B.8) by using the continuity equation (B.4), which results in

$$\frac{\partial\bar{u}}{\partial t} + \bar{u}\frac{\partial\bar{u}}{\partial x} + \bar{v}\frac{\partial\bar{u}}{\partial y} - f\bar{v} = -g\,\frac{\partial\zeta}{\partial x} - \frac{1}{\rho}\frac{\tau_x}{H+\zeta}$$

$$- \frac{1}{H+\zeta}\frac{\partial}{\partial x}\left[(H+\zeta)\overline{\hat{u}^2}\right] - \frac{1}{H+\zeta}\frac{\partial}{\partial y}\left[(H+\zeta)\overline{\hat{u}\hat{v}}\right]. \tag{B.9}$$

This is (6.14), except for the last two terms, which we ignored. They can be included in a parameterized form, which means that they are assumed to be somehow related to the depth-averaged variables. A common way to handle this is by casting them as diffusive terms, like

$$\overline{\hat{u}^2} = A\frac{\partial\bar{u}}{\partial x}, \qquad \overline{\hat{u}\hat{v}} = A\frac{\partial\bar{u}}{\partial y},$$

with a constant of diffusivity A. On that assumption, the system is closed.

References

Agnew, D. C. 2007. Earth tides. Pages 163–195 of: Herring, T. A. (ed), *Treatise on Geophysics, Vol. 3: Geodesy*. New York: Elsevier.

Arbic, B. K. 2005. Atmospheric forcing of the oceanic semidiurnal tide. *Geophys. Res. Lett.*, **32**(L02610).

Arbic, B. K., Mitrovica, J. X., MacAyeal, D. R., and Milne, G. A. 2008. On the factors behind large Labrador Sea tides during the last glacial cycle and the potential implications for Heinrich events. *Paleoceanogr.*, **23**(PA3211).

Baker, T. F. 1984. Tidal deformations of the Earth. *Sci. Prog. Oxf.*, **69**, 197–233.

Bartels, J. 1957. Gezeitenkräfte. Pages 734–774 of: Flügge, S., and Bartels, J. (eds), *Encyclopedia of Physics, Vol. XLVII: Geophysics II*. Berlin: Springer.

Beerens, S. P., Ridderinkhof, H., and Zimmerman, J. T. F. 1994. An analytical study of chaotic stirring in tidal areas. *Chaos, Solitons & Fractals*, **4**(6), 1011–1029.

Bills, B. G., and Ray, R. D. 1999. Lunar orbital evolution: A synthesis of recent results. *Geophys. Res. Lett.*, **26**(19), 3045–3048.

Blanco, V. M., and McCuskey, S. W. 1961. *Basic Physics of the Solar System*. Reading: Addison-Wesley.

Burchard, H., and Baumert, H. 1998. The formation of estuarine turbidity maxima due to density effects in the salt wedge: A hydrodynamic process study. *J. Phys. Oceanogr.*, **28**, 309–321.

Burchard, H., and Hetland, R. D. 2010. Quantifying the contributions of tidal straining and gravitational circulation to residual circulation in periodically stratified tidal estuaries. *J. Phys. Oceanogr.*, **40**, 1243–1262.

Carter, G. S., Fringer, O. B., and Zaron, E. D. 2012. Regional models of internal tides. *Oceanography*, **25**(2), 56–65.

Cartwright, D. E. 1978. Oceanic tides. *Int. Hydrogr. Rev.*, **LV**(2), 35–84.

Cartwright, D. E. 1999. *Tides: A Scientific History*. Cambridge: Cambridge University Press.

Cartwright, D. E., and Edden, A. C. 1973. Corrected tables of tidal harmonics. *Geophys. J. R. Astr. Soc.*, **33**, 253–264.

Cartwright, D. E., and Taylor, R. J. 1971. New computations of the tide-generating potential. *Geophys. J. R. Astr. Soc.*, **23**, 45–74.

Chapman, S., and Lindzen, R. S. 1970. *Atmospheric Tides: Thermal and Gravitational*. Dordrecht: Reidel.

Darwin, G. H. 1911. *The Tides and Kindred Phenomena in the Solar System*. 3rd edn. London: John Murray.

De Swart, H. E., and Zimmerman, J. T. F. 2009. Morphodynamics of tidal inlet systems. *Annu. Rev. Fluid Mech.*, **41**, 203–229.

Desai, S. D. 2002. Observing the pole tide with satellite altimetry. *J. Geophys. Res.*, **107**(C11), 3186.

Dijkstra, Y. M., Schuttelaars, H. M., and Burchard, H. 2017. Generation of exchange flows in estuaries by tidal and gravitational eddy viscosity-shear covariance (ESCO). *J. Geophys. Res.*, **122**, 4217–4237.

Doodson, A. T. 1921. The harmonic development of the tide-generating potential. *Proc. R. Soc. London, A*, **100**, 305–329.

Doodson, A. T. 1958. Oceanic tides. *Adv. Geophys.*, **5**, 117–152.

Doodson, A. T., and Warburg, H. D. 1941. *Admiralty Manual of Tides*. London: His Majesty's Stationary Office.

Duran-Matute, M., and Gerkema, T. 2015. Calculating residual flows through a multiple-inlet system: The conundrum of the tidal period. *Oc. Dyn.*, **65**, 1461–1475.

Duran-Matute, M., Gerkema, T., and Sassi, M. G. 2016. Quantifying the residual volume transport through a multiple inlet system in response to wind forcing: The case of the western Dutch Wadden Sea. *J. Geophys. Res.*, **121**, 8888–8903.

Egbert, G. D., and Ray, R. D. 2001. Estimates of M_2 tidal energy dissipation from TOPEX/Poseidon altimeter data. *J. Geophys. Res.*, **106**(C10), 22475–22502.

Egbert, G. D., and Ray, R. D. 2003a. Deviation of long-period tides from equilibrium: kinematics and geostrophy. *J. Phys. Oceanogr.*, **33**, 822–839.

Egbert, G. D., and Ray, R. D. 2003b. Semi-diurnal and diurnal tidal dissipation from TOPEX/Poseidon altimetry. *Geophys. Res. Lett.*, **30**(17), 1907.

Egbert, G. D., and Ray, R. D. 2017. Tidal prediction. *J. Mar. Res.*, **75**, 189–237.

Egbert, G. D., Ray, R. D., and Bills, B. G. 2004. Numerical modeling of the global semidiurnal tide in the present day and in the last glacial maximum. *J. Geophys. Res.*, **109**(C03003).

Ekman, M. 1993. A concise history of the theories of tides, precession-nutation and polar motion (from antiquity to 1950). *Surv. Geophys.*, **14**, 585–617.

Feistel, R., and Hagen, E. 1995. On the GIBBS thermodynamic potential of seawater. *Prog. Oceanogr.*, **36**, 249–327.

Fitzpatrick, R. 2012. *An Introduction to Celestial Mechanics*. Cambridge: Cambridge University Press.

Garrett, C. 1972. Tidal resonance in the Bay of Fundy and Gulf of Maine. *Nature*, **238**, 441–443.

Garrett, C., and Gerkema, T. 2007. On the body-force term in internal-tide generation. *J. Phys. Oceanogr.*, **37**, 2172–2175.

Garrett, C., and Greenberg, D. 1977. Predicting changes in tidal regime: The open boundary problem. *J. Phys. Oceanogr.*, **7**, 171–181.

Garrett, C., and Kunze, E. 2007. Internal tide generation in the deep ocean. *Annu. Rev. Fluid Mech.*, **39**, 57–87.

Garrett, C., and St. Laurent, L. 2002. Aspects of deep ocean mixing. *J. Oceanogr.*, **58**, 11–24.

Gerkema, T., and Gostiaux, L. 2012. A brief history of the Coriolis force. *Europhysics News*, **43**(2), 14–17.

Gerkema, T., and Shrira, V. I. 2005. Near-inertial waves in the ocean: Beyond the "traditional approximation." *J. Fluid Mech.*, **529**, 195–219.

Gerkema, T., and Van Haren, H. 2007. Internal tides and energy fluxes over Great Meteor Seamount. *Ocean Sci.*, **3**, 441–449.

Gerkema, T., Nauw, J. J., and Van der Hout, C. M. 2014. Measurements on the transport of suspended particulate matter in the Vlie Inlet. *Neth. J. Geosci.*, **93**(3), 95–105.

Geyer, W. R., and MacCready, P. 2014. The estuarine circulation. *Annu. Rev. Fluid Mech.*, **46**, 175–197.

Godin, G. 1972. *The Analysis of Tides*. Toronto: University of Toronto Press.

Goldstein, H. 1980. *Classical Mechanics*. 2nd edn. Reading: Addison-Wesley.

Gräwe, U., Flöser, G., Gerkema, T., Duran-Matute, M., Badewien, T. H., Schulz, E., and Burchard, H. 2016. A numerical model for the entire Wadden Sea: Skill assessment and analysis of hydrodynamics. *J. Geophys. Res.*, **121**, 5231–5251.

Green, J. A. M., and Nycander, J. 2013. A comparison of tidal conversion parameterizations for tidal models. *J. Phys. Oceanogr.*, **43**, 104–119.

Green, J. A. M., Huber, M., Waltham, D., Buzan, J., and Wells, M. 2017. Explicitly modelled deep-time tidal dissipation and its implication for Lunar history. *Earth Plan. Sci. Lett.*, **461**, 46–53.

Green, J. A. M., Molloy, J. L., Davies, H. S., and Duarte, J. C. 2018. Is there a tectonically driven supertidal cycle? *Geophys. Res. Lett.*, **45**, 3568–3576.

Guérin, O. 2004. *Tout Savoir sur les Marées*. Rennes: Editions Ouest-France.

Hansen, K. S. 1982. Secular effects of oceanic tidal dissipation on the moon's orbit and the earth's rotation. *Rev. Geophys. Space Phys.*, **20**(3), 457–480.

Harris, D. L. 1991. Reproducibility of the harmonic constants. Pages 753–770 of: Parker, B. B. (ed), *Tidal Hydrodynamics*. New York: Wiley.

Hendershott, M. C. 1981. Long waves and ocean tides. Pages 292–341 of: Warren, B. A., and Wunsch, C. (eds), *Evolution of Physical Oceanography*. Cambridge, Mass.: MIT Press.

Hibbert, A., Royston, S. J., Horsburgh, K. J., Leach, H., and Hisscott, A. 2015. An empirical approach to improving tidal predictions using recent real-time tide gauge data. *J. Oper. Oceanogr.*, **8**(1), 40–51.

Huthnance, J. M. 1973. Tidal current asymmetries over the Norfolk sandbanks. *Est. Coast. Mar. Sci.*, **1**, 89–99.

Jackson, C. R., Da Silva, J. C. B., and Jeans, G. 2012. The generation of nonlinear internal waves. *Oceanography*, **25**(2), 108–123.

Jackson, C. R., Da Silva, J. C. B., Jeans, G., Alpers, W., and Caruso, M. J. 2013. Nonlinear internal waves in synthetic aperture radar imagery. *Oceanography*, **26**(2), 68–79.

Jay, D. A., and Musiak, J. D. 1994. Particle trapping in estuarine tidal flows. *J. Geophys. Res.*, **99**(C10), 20445–20461.

Kantha, L. H., Stewart, J. S., and Desai, S. D. 1998. Long-period lunar fortnightly and monthly ocean tides. *J. Geophys. Res.*, **103**(C6), 12639–12647.

Krauss, W. 1966. *Interne Wellen*. Berlin: Gebrüder Borntraeger.

Lam, F. P. A. 1999. Shelf waves with diurnal tidal frequency at the Greenland shelf edge. *Deep-Sea Res. I*, **46**, 895–923.

Le Provost, C. 1991. Generation of overtides and compound tides (review). Pages 269–295 of: Parker, B. B. (ed), *Tidal Hydrodynamics*. New York: Wiley.

LeBlond, P. H., and Mysak, L. A. 1978. *Waves in the Ocean*. Amsterdam: Elsevier.

Lin, C. C., and Segel, L. A. 1974. *Mathematics Applied to Deterministic Problems in the Natural Sciences*. New York: Macmillan.

Longuet-Higgins, M. S. 1968a. The eigenfunctions of Laplace's Tidal Equations over a sphere. *Phil. Trans. R. Soc. London, A*, **262**(1132), 511–607.

Longuet-Higgins, M. S. 1968b. On the trapping of waves along a discontinuity of depth in a rotating ocean. *J. Fluid Mech.*, **31**(3), 417–434.

Lyard, F. H., and Le Provost, C. 1997. Energy budget of the tidal hydrodynamic model FES94.1. *Geophys. Res. Lett.*, **24**(6), 687–690.

Maas, L. R. M., and Van Haren, J. J. M. 1987. Observations on the vertical structure of tidal and inertial currents in the central North Sea. *J. Mar. Res.*, **45**, 293–318.

MacCready, P., and Geyer, W. R. 2010. Advances in estuarine physics. *Annu. Rev. Mar. Sci.*, **2**, 35–58.

Morozov, E. G. 2018. *Oceanic Internal Tides: Observations, Analysis and Modeling: A Global View*. Berlin: Springer.

Munk, W., and Bills, B. 2007. Tides and the climate: Some speculations. *J. Phys. Oceanogr.*, **37**, 135–147.

Munk, W., and Wunsch, C. 1997. The Moon, of course. *Oceanography*, **10**(3), 132–134.

Munk, W., and Wunsch, C. 1998. Abyssal recipes II: Energetics of tidal and wind mixing. *Deep-Sea Res. I*, **45**(3), 1977–2010.

Nicolas, G. 1995. *Introduction to Nonlinear Science*. Cambridge: Cambridge University Press.

Officer, C. B. 1976. *Physical Oceanography of Estuaries (And Associated Coastal Waters)*. New York: Wiley.

Oonishi, Y., and Kunishi, H. 1979. Water exchange between adjacent vortices under an additional oscillatory flow. *J. Oceanogr. Soc. Japan*, **35**, 136–140.

Open University. 1989. *Waves, Tides and Shallow-Water Processes*. Oxford: Pergamon.

Palmer, J. D. 1996. Time, tide and the living clocks of marine organisms. *Am. Sci.*, **84**, 570–578.

Parker, B. B. 1991. The relative importance of the various nonlinear mechanisms in a wide range of tidal interactions (review). Pages 237–268 of: Parker, B. B. (ed), *Tidal Hydrodynamics*. New York: Wiley.

Parker, B. B. 2011. The tide predictions for D-Day. *Phys. Today*, **64**(9), 35–40.

Pawlowicz, R., Beardsley, B., and Lentz, S. 2002. Classical tidal harmonic analysis including error estimates in MATLAB using T_TIDE. *Computers & Geosci.*, **28**, 929–937.

Platzman, G. W. 1984. Normal modes in the world ocean. Part IV: Synthesis of diurnal and semidiurnal tides. *J. Phys. Oceanogr.*, **14**, 1532–1550.

Prandle, D. 1982. The vertical structure of tidal currents and other oscillatory flows. *Cont. Shelf Res.*, **1**(2), 191–207.

Proudman, J. 1953. *Dynamical Oceanography*. London: Methuen.

Pugh, D., and Woodworth, P. 2014. *Sea-Level Science: Understanding Tides, Surges, Tsunamis and Mean Sea-Level Changes*. Cambridge: Cambridge University Press.

Ray, R. D. 1998. Ocean self-attraction and loading in numerical tidal models. *Mar. Geod.*, **21**(3), 181–192.

Ray, R. D. 2007. Propagation of the overtide M_4 through the deep Atlantic Ocean. *Geophys. Res. Lett.*, **34**(L21602).

Ray, R. D., and Cartwright, D. E. 2007. Times of peak astronomical tides. *Geophys. J. Int.*, **168**, 999–1004.

Ray, R. D., and Egbert, G. D. 2004. The global S_1 tide. *J. Phys. Oceanogr.*, **34**, 1922–1935.

Ray, R. D., Eanes, R. J., and Lemoine, F. G. 2001. Constraints on energy dissipation in the Earth's body tide from satellite tracking and altimetry. *Geophys. J. Int.*, **144**, 471–480.

Richards, E. G. 1998. *Mapping Time: The Calendar and Its History*. Oxford: Oxford University Press.

Ridderinkhof, H., and Zimmerman, J. T. F. 1992. Chaotic stirring in a tidal system. *Science*, **258**, 1107–1111.

Roberts, J. 1975. *Internal Gravity Waves in the Ocean*. New York: Marcel Dekker.

Robinson, I. S. 1983. Tidally induced residual flows. Pages 321–356 of: Johns, B. (ed), *Physical Oceanography of Coastal and Shelf Seas*. New York: Elsevier.

Roos, P. C., and Schuttelaars, H. M. 2011. Influence of topography on tide propagation and amplification in semi-enclosed basins. *Oc. Dyn.*, **61**, 21–38.

Russell, H. N., Dugan, R. S., and Stuart, J. Q. 1926. *Astronomy: A Revision of Young's Manual of Astronomy. Volume I: The Solar System*. Boston: Ginn and Company.

Sanchez, B. V. 2008. Normal Modes of the Global Oceans: A Review. *Mar. Geod.*, **31**(3), 181–212.

Schureman, P. 1940. *Manual of Harmonic Analysis and Prediction of Tides*. Washington: U. S. Coast and Geodetic Survey, Spec. Publ. No. 98.

Serrin, J. 1959. Mathematical principles of classical fluid mechanics. Pages 125–263 of: Flügge, S., and Truesdell, C. (eds), *Encyclopedia of Physics, Vol. VIII: Fluid Dynamics I*. Berlin: Springer.

Shapiro, G. I. 2011. Effect of tidal stream power generation on the region-wide circulation in a shallow sea. *Ocean Sci.*, **7**, 165–174.

Simmons, H. L., Jayne, S. R., St. Laurent, L. C., and Weaver, A. J. 2004. Tidally driven mixing in a numerical model of the ocean general circulation. *Ocean Model.*, **6**, 245–263.

Simpson, J. H., Brown, J., Matthews, J., and Allen, G. 1990. Tidal straining, density currents, and stirring in the control of estuarine stratification. *Estuaries and Coasts*, **13**(2), 125–132.

Smith, W. H. F., and Sandwell, D. T. 1997. Global sea floor topography from satellite altimetry and ship depth soundings. *Science*, **277**, 1956–1962.

Spencer, J. 2011. Watery Enceladus. *Phys. Today*, **64**(11), 38–44.

Stammer, D., et al. 2014. Accuracy assessment of global barotropic ocean tide models. *Rev. Geophys.*, **52**, 243–282.

Stephenson, F. R. 2003. Historical eclipses and Earth's rotation. *Astronomy & Geophysics*, **44**, 2.22–2.27.

Sutherland, G., Garrett, C., and Foreman, M. 2005. Tidal resonance in Juan de Fuca Strait and the Strait of Georgia. *J. Phys. Oceanogr.*, **35**, 1279–1286.

Tamisiea, M. E., and Mitrovica, J. X. 2011. The moving boundaries of sea level change: Understanding the origins of geographic variability. *Oceanography*, **24**(2), 24–39.

Taylor, G. I. 1922. Tidal oscillations in gulfs and rectangular basins. *Proc. London Math. Soc.*, **20**, 148–181.

Tessmar-Raible, K., Raible, F., and Arboleda, E. 2011. Another place, another timer: Marine species and the rhythms of life. *Bioessays*, **33**, 165–172.

Toksöz, M. N., Goins, N. R., and Cheng, C. H. 1977. Moonquakes: Mechanisms and relation to tidal stresses. *Nature*, **196**(4293), 979–981.

Uncles, R. J. 2002. Estuarine physical processes research: Some recent studies and progress. *Estuarine Coast. Shelf Sci.*, **55**, 829–856.

Vallis, G. K. 2006. *Atmospheric and Oceanic Fluid Dynamics: Fundamentals and Large-Scale Circulation*. Cambridge: Cambridge University Press.

Van de Kreeke, J., and Brouwer, R. L. 2017. *Tidal Inlets: Hydrodynamics and Morphodynamics*. Cambridge: Cambridge University Press.

Van der Hout, C. M., Witbaard, R., Bergman, M. J. N., Duineveld, G. C. A., Rozemeijer, M. J. C., and Gerkema, T. 2017. The dynamics of suspended particulate matter (SPM) and chlorophyll-*a* from intratidal to annual time scales in a coastal turbidity maximum. *J. Sea Res.*, **127**, 105–118.

Van Haren, H., and Gostiaux, L. 2009. High-resolution open-ocean temperature spectra. *J. Geophys. Res.*, **114**(C05005), 1–14.

Van Haren, H., and Gostiaux, L. 2012. Energy release through internal wave breaking. *Oceanography*, **25**(2), 124–131.

Van Veen, J. 1937. *Velocities in a Vertical Line of a Stream*. Rapporten en mededeelingen van den Rijkswaterstaat, No. 29. 's-Gravenhage: Alg. Landsdrukkerij.

Visser, A. W., Souza, A. J., Hessner, K., and Simpson, J. H. 1994. The effect of stratification on tidal current profiles in a region of freshwater influence. *Oceanologica Acta*, **17**(4), 369–381.

Vlasenko, V., Stashchuk, N., and Hutter, K. 2005. *Baroclinic Tides: Theoretical Modeling and Observational Evidence*. Cambridge: Cambridge University Press.

Wells, M. R., Allison, P. A., Piggott, M. D., Pain, C. C., Hampson, G. J., and de Oliveira, C. R. E. 2005. Large sea, small tides: the Late Carboniferous seaway of NW Europe. *J. Geol. Soc. London*, **162**, 417–420.

Williams, G. E. 2000. Geological constraints on the Precambrian history of Earth's rotation and the Moon's orbit. *Rev. Geophys.*, **38**(1), 37–59.

Woodworth, P. L. 2012. A note on the nodal tide in sea level records. *J. Coastal Res.*, **28**(2), 316–323.

Zimmerman, J. T. F. 1981. Dynamics, diffusion and geomorphological significance of tidal residual eddies. *Nature*, **290**(5807), 549–555.

Zimmerman, J. T. F. 1982. On the Lorentz linearization of a quadratically damped forced oscillator. *Phys. Lett.*, **89A**(3), 123–124.

Zimmerman, J. T. F. 1986. The tidal whirlpool: a review of horizontal dispersion by tidal and residual currents. *Neth. J. Sea Res.*, **20**(2/3), 133–154.

Zimmerman, J. T. F. 1993. *Cooscillation*. Lecture notes R93-8, IMAU, Utrecht University.

Index